Corinne Martin · Thilo von Pape (Eds.)

Images in Mobile Communication

VS RESEARCH

Corinne Martin
Thilo von Pape (Eds.)

Images in Mobile Communication

New Content, New Uses,
New Perspectives

VS RESEARCH

Bibliographic information published by the Deutsche Nationalbibliothek
The Deutsche Nationalbibliothek lists this publication in the Deutsche Nationalbibliografie;
detailed bibliographic data are available in the Internet at http://dnb.d-nb.de.

Published with financial support from the Paul Verlaine University of Metz (France) and its Centre for Research on Mediations (CREM), the Région Lorraine (France), the Urban Community of Metz Métropole (France), the Fédération française des télécoms (FFT) and InTech SA (Grand Duchy of Luxembourg).

1st Edition 2012

Editorial Office: Verena Metzger | Britta Göhrisch-Radmacher

VS Verlag für Sozialwissenschaften is a brand of Springer Fachmedien.
Springer Fachmedien is part of Springer Science+Business Media.
www.vs-verlag.de

Cover design: KünkelLopka Medienentwicklung, Heidelberg
Printed on acid-free paper
Printed in Germany

ISBN 978-3-531-17992-6

Contents

Introduction

Thilo von Pape and Corinne Martin

In the beginning was the word, and the word was through GSM, and GSM was voice only[1]. Despite the first generation of cellular networks in Japan, North America and Europe beginning in the early 1980s, mobile communication remained a marginal phenomenon for ten years until the emergence of second generation networks based on the digital standard of the "Global System for Mobile Communications" (GSM). What followed was a textbook example of the successful diffusion of innovations, starting with almost no adopters in 1992 to one subscription per European citizen in 2006 (Eurostat, 2006). It is also the evolution of a new medium, from a channel for the human voice to a universal means of communication. We argue that the advent of images is a key moment in the evolution of mobile communication. Here we outline how this book investigates this period throughout the subsequent parts and chapters.

1   From Voice Only to Written Text

When mobile phone adoption rapidly increased in the late 1990s, it was due not only to the dynamics inherent to diffusion. Rather, the evolution also benefited from the new short messaging service (SMS), which was embedded into the GSM standard and the software of existing phones. Embraced with much enthusiasm, particularly by adolescent phone users (cf. Ling, 2000), SMS accompanied and flourished with mobile communication in general. Communication, which previously could only be said and heard, was now visual. Some observers tend to relativize this advance, arguing that the language used in SMS messages is phonetic and therefore not anchored in the written tradition (Soffer, 2010), whereas others see it as a milestone in a universal comeback of the written language. In his social ontology of the mobile phone, Maurizio Ferraris (2006, p. 45) points out "we cannot agree with those who say that what we have here is a creolization, in which the written becomes a variant of the spoken", insisting that the mobile phone "is not [...] an instrument for talking but an instrument for

---

[1] Loosely based on the Gospel of John, 1:1.

writing" (Ferraris, 2006, p. 48). Consequently, Ferraris traces the genealogy of mobile phones not to Thomas Edison, but rather to the Sumerans' tablets. While this argument appeared to be surprising in 2006, it has gained legitimacy as the evolution of smart phones produced a new class of tablet devices, thus closing the circle.

If we assume that the introduction of short messages establishes a new trajectory of ancestry in the evolution of mobile communication, then it can be argued that this led to new uses and meanings of the technology, which were soon associated with mobile communication in general. As Goggin and Hjorth (2009, p. 3) claim, "text messaging took on a life of its own – spawning a career in technology that quickly moved beyond mere signal or data to direct suturing into youth culture, interactive television, mobile commerce, and so on". The beginning of this evolution is well documented. The "perpetual contact" that is now possible due to mobile communication (Katz & Aakhus, 2002), was extended through the asynchronous character of SMS communication and the opportunities it offered in order to communicate with friends and family even when they were not available. If the phone was a popular tool for "micro-coordination" in everyday life and the "hyper-coordination" of social relations (Ling & Yttri, 2002), these activities were reinforced by the low cost and convenience of short messages. Thus, teenagers could afford not only to arrange their leisure activities through SMS dialogues, but also to affirm and enhance their social relationships through sending symbolically charged messages to their friends in what Taylor and Harper (2002) compare to age-old gift-giving practices. In the course of this appropriation, mobile communication turned out to be more than merely words, as users applied a reinvention of letters to form smiley faces and other emoticons. Their seemingly inevitable presence in text messages today is seen by some – with reference to the school of Palo Alto – as an analogy to the unavoidable non-verbal elements in face-to-face communication (Rivière, 2002; Watzlawick et al., 1967).

2   From Words to Images

The success of emoticons foreshadowed a second major advancement in the qualitative evolution of mobile communication, which is at the heart of this book: the advent of images. Evolving from emoticons to black and white graphics, photos and videos, images on mobile phones soon improved in quality (e.g., the objectives of the camera, the size of the screen, resolution and the transmission bandwidths). This evolution ultimately permitted integration of the Internet into mobile communication and fulfilled the "convergence" between informatics, mass media and telecommunications for mobile end devices. The Internet experience demands images in convincing quality. This lesson could be learned from the failure of the "light" WAP format (Joakar & Fish, 2006), and Apple applied this lesson in the development of the first iPhone, which gave users an almost uncompromised experience of the Internet (West & Mace, 2010).

Now that the number of users with mobile Internet access is exploding, mobile phones are no longer just devices for telecommunication but are "*becoming media*" (Goggin & Hjorth, 2009, p. 3). While some researchers are exploring whether this shift will mark the beginning of a new era "after the mobile phone" (Hartmann, Rössler, & Höflich, 2008), it can be assumed that this is a very disruptive moment in the evolution of mobile communication.

Just as the arrival of text allowed for completely new meanings and uses of mobile communication, we argue that the arrival of images is a key moment that may determine the further development of mobile communication for a considerable amount of time. Enriching the evolution of mobile communication now is the wide visual tradition of popular culture and mass media. This raises questions about how to analyze these processes and the specific role images play in it.

3 "Thou Shalt Not Make unto Thee Any Graven Image"

A widespread choice for the study of images is to consider them as signs in the tradition of semiotics, as explored by Roland Barthes (1981). Building on Barthes, Rivière (2005) suggested that snapshots produced by mobile phones break with the relation to time that is characteristic of traditional photographs, through the idea that photographs testify that "that has been." While customary photographs juxtapose present reality with the past by both authenticating the latter and representing it – often creating a nostalgic emotion – mobile snapshots are mainly embedded in the present and develop their meaning as such. Whether they are immediately watched on a camera's display or shared wirelessly with others, they evoke an instantaneous sense of "being together". Thus, they introduce a new imaginary and visual language that is also embedded in the present. However, just as written words are more than the written form of oral language, some argue that images also go beyond what can be reduced to language. One such argument that underlies the difference between words and images can be found in the bible and history of the church. Consider, for example, the Second Commandment, "Thou shalt not make unto thee any graven image, or any likeness *of any thing* that *is* in heaven above, or that *is* in the earth beneath, or that *is* in the water under the earth" (King James Bible, edited by Carroll & Pricket,

2008). Not only does the existence and prominence of this interdiction among the commandments underline the power attributed to images (as opposed to text), but its importance is also confirmed by the course of church history. Marie-José Mondzain (2005) makes this point in her analysis of the byzantine roots of our contemporary understanding and use of images. She refers to the iconoclasm under Constantine V in the VIIIth and IXth century, and more precisely to the patriarch Nikephoros who defended images, to explain to what extent they are completely different from text: "When the image is operative, it does something that the speech does not. [...] Nikephoros reminds us that the image is a Gospel and that there is a perfect equivalence between the scriptural message and the iconic message. Nonetheless, he never loses sight of the fact that teaching and persuading by means of the icon are superior to hearing as a result of the speed with which they operate, and their emotional effectiveness" (Mondzain, 2005, p. 6). Through their emotional efficiency, images are vectors of power. Michel Foucault expresses the same idea of untranslatability in his analysis of the painting *Las Meninas* (Vélasquez) in *The Order of Things: An Archaeology of the Human Sciences* (1994): "it is not that words are imperfect, or that, when confronted by the visible, they prove insuperably inadequate. Neither can be reduced to the other's terms: it is in vain that we say what we see; what we see never resides in what we say" (Foucault, 1994, p. 10).

Images have functions beyond the symbolization of semiotics; according to the psychologist Serge Tisseron (2007), the image transforms the spectator and is transformed by him/her, which provokes the question of "truthfulness" in the digital age. Secondly, the image works as an "envelope" for the viewer, who can then "inhabit" the image. The contemplation of images and access to the imaginary is the origin of the happiness we find in an image. It is through this process that the famous phrase "I was there" is expressed, which implicitly accompanies the exchange of photos/videos among friends.

4   Studying Images in Mobile Communication

How can we analyze the role of images in mobile communication, if we want to go beyond a perspective that considers them as signs? One approach for positively understanding the significance of images is "visual studies", which was created in the 1980s as a confluence of art history, esthetics, history of ideas and cultural studies (Mitchell, 1986). While these disciplines may appear distant from the questions confronting mobile communication today, there are actually some parallels. For example, the work of Marie-José Mondzain (2005) on iconoclasm and its meaning for our contemporary understanding of images suggests a connection between mobile communication and art history. Mondzain links images not only to religious power but also to economic power, which is much more salient in the struggle today regarding the future of mobile communication. As Mondzain claims, "The question of the economy cannot be separated from the question of the image itself" (Mondzain, 2005, p. 3). This confers a very particular power to the image: "It is precisely because the icon is endowed with a power specific to it that it mattered so much to the emperor to deprive the church of it and to reserve for himself its exclusive rights and benefits" (Mondzain, 2005, p. 6). This power acquired by the image appears today in the age of digital devices.

The advent of images on the screens of mobile phones can crystallize and reveal the enormous economic stakes of the so-called convergence between cultural industries (e.g., music, press, information, editing, cinema and audiovisual media) and the industries of communication and information (Bouquillion, 2008). The mobile phone may become the mass medium that fosters the diffusion of visual content: from the photo to television, games to graphic novels, and other content available online. The entry of new actors from the communication industry means that the cards are reshuffled and distributed to these new actors. If their strategy aims mainly at differentiating their offer from the competition (and not to transform themselves into producers of content or holders of exclusive rights), large actors such as the mobile operators will compete with content producers in order to keep the position they own through their direct access to the consumers (Bouquillion, 2008). These large actors of communication tend to become infomediaries (intermediaries between producers of content and consumers/users) and acquire meta-positions as they enter the content business (Bouquillion, 2008). The most recent example of this is Apple's strategy to bypass the traditional relationship between providers of content (such as publishers or record companies) and consumers by establishing direct and exclusive links of users to their iPhones (West & Mace, 2010) and iPads.

5   Outlook: Parts and Chapters

Thus far, it can be concluded that the stakes as well as the rules of the game and the players change when images are involved. This follows historical trends and seemingly applies to the current evolution in mobile communication as well. Given the wide scope of these changes, this book deliberately explores only three axes, each of which is addressed in a part containing three chapters.

The first part of the book, titled *"The Production of Photography in the Users' Hands?"* addresses the consequences of utilizing mobile phones as cameras.

In the first chapter of this part, *"Images in Mobile Uses: A New "Middle-brow" Art?"*, Anne Jarrigeon draws on Bourdieu's research on the "social uses of photography" in order to examine whether the mobile phone facilitates a new multimedia "middle-brow" art. She first explores what new types of images are created and produced with mobile phones. Relying on an analysis of the continuities and discontinuities provoked by new visual uses of the cell phone, this approach attempts to uncover the aesthetic template common to a large number of mobile productions. This approach is based on studies of various cases, such as family photos, citizen journalism, everyday life ethnography, spectacular performance and reflexive use. Several hypotheses about the way mobile phone pictures fit into our historical visual culture have been proposed and these stimulate further questions on how these images affect our sense of representation as well as our social interactions. Everyday snapshots also play an important role in the work of Iren Schulz titled *"Visual Mobile Phone Content and Developmental Challenges: The Mediatization of Social Relationships in Adolescence"*. However, this chapter has both a broader and tighter focus as it explores the topics and types of any visual content that can be found on adolescents' mobile phones including the context in which it is exchanged. Schulz links this content to strategies of participation in adolescent relationships. Groups of peers hence appear as dynamic social networks of negotiated meanings. Ranging from everyday-life situations via favorite media content to sexuality and violence, visual content and its use reveal meaningful differences beyond gender and education. Schulz's empirical findings are based on pictures, photos and video clips that are examined alongside the in-depth interviews of ten adolescents. The final chapter of the first part, *"Celebration and Concern: Digitization, Camera Phones and the Citizen-Photographer"* by Frank Möller, addresses the changes created by digitization to the techniques of producing, altering and disseminating images and discusses the social and political consequences of these new possibilities. The photos produced by simple citizens may help construct an international visual/virtual community, but they are not always welcome due to the difficulty of controlling them and their ability to undermine authority. Möller analyzes citizen-photography by discussing the possibilities that digitization, especially the use of camera phones, offers to marginalized groups of people to exert agency by becoming photographers. The author chose photographs of the 2005 bombings on the London Underground and of the former Prime Minister Tony Blair. Furthermore, he focuses on both photojournalism and irony in the digital age with a discussion on the concept of "sousveillance".

The second part of the volume covers *"Strategies and Tactics at the Advent of Mobile Images"*. It comprises three empirical studies on the ways users and producers prepare and react to the potential of mobile communication receiving multimedia content and multimedia content becoming mobile. Cornelia Wolf and Ralf Hohlfeld consider this question from the perspective of those actors providing journalistic content by administering an editorial survey among the news desks of German radio stations, TV channels, as well as newspapers and newsmagazines with local regional and national coverage. If their title *"Revolution in Journalism? Mobile Devices as a New Means of Publishing"* raises the question of how profound the upcoming changes to journalism will actually be, the data confirm that the producers expect them to have a large impact on news production and the very structure of journalistic work and organizations. Perhaps the least predictable variable in the producers' expectations are the users and their acceptance and appropriation of mobile services. Both Thilo von Pape and Veronika Karnowski and Julien Figeac deal with these questions. Unlike the news desks and device manufacturers, users generally do not create the content, devices or environment in which they use mobile media. Rather, they try to maximize usage of the given technology by applying what de Certeau (1984) would describe as using their own tactics on the foreign "territory". In the chapter *"Which Place for Mobile Television in Everyday Life? Evidence from a Panel Study"*, Von Pape and Karnowski explore, how users integrate a mobile television service into their everyday life over the time span of 12 weeks. Based on a mobile experience sampling methodology, their findings question the optimistic visions that often accompany the idea of mobile television. Yet they also suggest how this struggling technology may still be successful among users and find a place in their everyday routines. Whereas these findings draw on mostly quantitative data, Julien Figeac presents a purely qualitative study on the same object of research, using a similarly innovative methodology. Titled *"The Appropriation of Mobile TV through Television Preferences and Communication Networks"*, this chapter provides ethnographic observations of Parisian Mobile TV subscribers. It relies on recordings from camera glasses that users wore during a week of mobile TV usage in addition to in-depth interviews. Thus radically, and literally, taking the users' perspective, Figeac is able to see the problems that users are confronting regarding network availability. Insights into the actual situational context uncover not only preferences in terms of content, but also the conditions of reception. Carefully analyzing users reflexivity toward both content and conditions, Figeac comes up with the notion of "opportunist attachment" to describe the users' relation with mobile television today. Given that the utilizations of technological opportunities are heavily restricted, the tactics with which users seize these opportunities for specific forms of everyday use provides a more in-depth understanding than any survey among users, let alone the usage scenarios developed by the producers.

The third part of the book, *"Images and Representations of Mobile Communication"*, examines the image and social representations of mobile communication itself. The first chapter, *"Images and Representations of the Mobile Internet"* by Corinne Martin, addresses questions about the producers' strategies as discussed in the previous part of the volume, but then applies them to the question of how various social actors create different images of the emerging mobile Internet. The first image of the mobile Internet user, as constructed by various audience measurements in France, appears as a blurred and embellished picture that is still under construction. The second image is produced in the public space by the aggressive strategies of network operators and handset manufacturers, as will be illustrated in a case study of the iPhone. Third, Martin utilizes the concept of social representation in order to analyze the users' representation of the mobile Internet in relation to the computer. Mixed methodology is applied, utilizing analysis of statistics of use and experts' forecasts, questionnaires and in-depth interviews among young adults. In comparison with this chapter on the very specific application of the mobile Internet, the last two chapters deal with the more general question of how mobile communication as a whole is socially constructed. More precisely, these chapters are dedicated to the role of images for the construction of mobile communication in the context of advertisement and TV series. Based on the assumption that actors in television series may serve as role models for their viewers, Veronika Karnowski analyzes how mobile phone use is represented in TV serials broadcast in Germany. The findings are presented in the chapter *"Symbolic Models of Mobile Phone Appropriation: A Content Analysis of TV Serials"*. These longitudinal insights give a unique impression of how fictional audiovisual media have pictured the evolving innovation of mobile communication with respect to its various dimensions (such as pragmatic use, symbolic use, restrictions). Another visual reflection of mobile communication, even more directly aimed at reconstructing the evolving innovation by influencing its users, is the representation in graphical advertisements. Such advertisements, appearing in the Spanish journals *El País* and *El Mundo,* are the basis of Miguel Ángel Ojeda's chapter *"The Image of Youth in Mobile Phone Advertising"*. In his analysis, Ojeda attempts to shape the image of mobile communication with the characteristics of youth culture. He offers a more subtle effort to frame youth culture as a virtual phenomenon, taking place in the digital space of mobile social networking.

While these three parts are distinct due to the specificity of their questions, they also constitute three levels of time and reflection, leading the reader from common uses and meanings of mobile images today (part 1) via expectations and pilot studies of their future production and appropriation (part 2), to representations and visual reflections of mobile communication over longer periods of time (part 3).

6   Acknowledgements

The editors of this book gratefully acknowledge the intellectual, practical and financial support they have received from both scholars and institutions. The members of the international editorial board have reviewed all proposals and given important suggestions, thus helping to find the best possible contributions in addition to improving the quality of the work. Our thanks go to Richard Ling (IT University of Copenhagen, Denmark), Christian Licoppe (Telecom Paris Tech, France), Francis Jauréguiberry (University of Pau, France), Sebastian Schnorf (Harvard University, United States), Luc Massou, Brigitte Simonnot and Sébastien Genvo (Paul Verlaine University of Metz, France). We also thank Elisabeth Günther (Hohenheim University in Stuttgart, Germany) for turning the various contributions into one homogeneous format ready to print. Finally, we thank the Paul Verlaine University of Metz and its Centre for Research on Mediations (CREM), the *Région Lorraine* (France), the Urban Community of *Metz Métropole* (France), the *Fédération française des télécoms* (FFT) and *InTech SA* (Grand Duchy of Luxembourg) whose financial support made this publication possible.

References

Barthes, R. (1981). *Camera Lucida: Reflections on Photography*. New York: Hill and Wang. (Original work published 1980)

Bourdieu, P. (1990). *Photography: A Middle-brow Art*. Stanford: Stanford University Press. (Original work published 1965)

Carroll, R. & Prickett, S. (2008). *The Bible: Authorized King James Version* (Oxford World's Classics). New York: Oxford University Press.

Certeau, M. de (1984). *The Practice of Everyday Life*. Berkeley: University of California Press. (Original work published 1980)

Eurostat (2006). *Mobile phone subscriptions (per 100 inhabitants)*. Available online: http://epp.eurostat.ec.europa.eu/tgm/table.do?tab=table&init=1&plugin=1&language=en&pcode=tin00060 [June 6, 2011].

Foucault, M. (1994). *The Order of Things: An Archaeology of the Human Sciences*. New York: Vintage Books. (Original work published 1966)

Ferraris, M. (2006). Where Are You? Mobile Ontology. In K. Nyíri (Ed.), *Mobile Understanding: The Epistemology of Ubiquitous Communication* (pp. 41-53). Wien: Passagen Verlag.

Goggin, G. & Hjorth, L. (2009). *Mobile Technologies: From Telecommunications to Media*. New York, Abingdon: Routledge.

Hartmann, M., Rössler, P., & Höflich, J. (2008). *After the Mobile Phone? Social Changes and the Development of Mobile Communication*. Berlin: Frank & Timme.

Jaokar, A. & Fish, T. (2006). *Mobile Web 2.0: The Innovator's Guide to Developing and Marketing Next Generation Wireless/Mobile Applications*. London: Futuretext.

Katz, J. E. & Aakhus, M. (Eds.) (2002). *Perpetual Contact: Mobile Communication, Private Talk, Public Performance*. Cambridge: Cambridge University Press.

Licoppe, C. (2002). Sociabilité et technologies de communication: Deux modalités d'entretien des liens interpersonnels dans le contexte du déploiement des dispositifs de communication mobiles. *Réseaux, 20*(112-113), 171-210.

Ling, R. & Yttri, B. (2002). Hyper-coordination via mobile phones in Norway. In J. E. Katz and M. Aakhus (Eds.), *Perpetual Contact: Mobile Communication, Private Talk, Public Performance* (pp. 139-169). Cambridge: Cambridge University Press.

Ling, R. (2010). Texting as a life-phase medium. *Journal of Computer-Mediated Communication, 15*, 277-292.

Ling, R. (2000). "We will be reached": The use of mobile telephony among Norwegian youth. *Information Technology & People, 13*(2), 102-120.

Meeker, G. M. (2009). *The Mobile Internet Report*. Morgan Stanley. Available online: http://www.morganstanley.com/institutional/techresearch/mobile_Internet_report12 2009.html [January 14, 2011].

Mitchell, W. J. T. (1986). *Iconology: Image, Text, Ideology*. Chicago: The University of Chicago Press.

Mondzain, M.-J. (2005). *Image, Icon, Economy: The Byzantine Origins of the Contemporary Imaginary*. Standford: Standford University Press. (Original work published 1996)

Pequignot, B. (2008). A-t-on besoin des images ? In F. Gaudez (Ed.), *Les arts moyens aujourd'hui* (pp. 15-29). Paris: L'Harmattan.

Rivière, C.-A. (2002). La pratique du mini-message: une double stratégie d'extériorisation et de retrait de l'intimité dans les interactions quotidiennes. *Réseaux, 20*(112-113), 139-168.

Rivière, C.-A. (2005). Mobile Camera Phones: A New Form of "Being Together" in Daily Interpersonal Communication. In R. Ling & P. E. Pedersen (Eds.), *Mobile Communications: Re-negotiation of the Social Sphere* (pp. 167-185). London: Springer.

Soffer, O. (2010). "Silent Orality": Toward a Conceptualization of the Digital Oral Features in CMC and SMS Texts. *Communication Theory, 20*(4), 387-404.

Taylor, A. S. & Harper, R. (2002, April). Age-old practices in the 'New World': A study of gift-giving between teenage mobile phone users. Paper presented at the Conference on Human Factors and Computing systems (CHI 2002), Minneapolis, Minnesota. Available online: http://research.microsoft.com/en-us/um/people/ast/files/CHI_2002.pdf [January 14, 2011].

Tisseron, S. (1996). *Le bonheur dans l'image*. Le Plessis-Robinson: Institut Synthélabo pour le progrès de la connaissance.

Tisseron, S. (2005). *Psychanalyse de l'image: Des premiers traits au virtuel.* Paris: Dunod.

Watzlawick, P., Beavin, J.-H., & Jackson, D. (1967). *Pragmatics of Human Communication: A Study of Interactional Patterns, Pathologies, and Paradoxes.* New York: Norton & Company.

West, J. & Mace, M. (2010). Browsing as the killer app: Explaining the rapid success of Apple's iPhone. *Telecommunications Policy, 34*(5-6), 270-286.

Wei, R. (2008). Motivations for using the mobile phone for mass communications and entertainment. *Telematics and Informatics, 25*, 36-46.

Part I

The Production of Photography in the Users' Hands?

Images in Mobile Uses: A New "Middle-brow Art"?[1]

Anne Jarrigeon

1 Introduction

Today, the use of the multimedia functions of mobile phones is far from marginal and is not limited to the younger generation. In France, among the "extended uses" of the telephone (Licoppe & Zouinar, 2009), video and photography are progressing the most rapidly.[2] From the unique experiments on the first day when one explores a new telephone, to hastily taken photographs of one's friends and family, to advanced use of these audiovisual devices by artists, the use of mobile phones' image-making capabilities has begun to redefine common practices dealing with images and sound. These effects extend beyond the field of interpersonal relationships and interactions upon which many researchers choose to concentrate (Riviere, 2005, 2006; Koskinen, 2007).[3] "There are indicators of practices of picture taking and sharing that differ both from the uses of the stand-alone camera and the kinds of social sharing that happened via mobile phone communication", write Okabe and Ito (2006, p. 79).

In fact, artistic practices are developing alongside "vulgar" uses. In this way, the mobile phone stands out from the multitude of objects that communicate or aid in communication, such as landline telephones, computers, or PDAs. It is also making a place for itself in the history of photographic devices such as the disposable camera, the Polaroid, and the digital camera or, in the realm of animated images, portable super-8 and video cameras. These audiovisual practices are new, yet they maintain a dialogue with existing devices. It is therefore important to examine them closely from both the perspective of developing amateur practices, which build upon family photography and videos, and the pre-

---

[1]   Bourdieu et al., 1965
[2]   The latest survey commissioned by the AFOM (French Association of Mobile Operators) from TNS-Sofres (TNS-Sofres, 2009). Available: http://www.archive.afom.fr/v4/STATIC/ documents/OBS_2009_AFOM_TNS_Sofres.pdf [November 15, 2010].
[3]   For example, Ilpo Koskinen employs conversation analysis ethnomethodology and considers "mobile multimedia as interaction" (Koskinen, 2007, p. 33).

mise of an artistic aesthetic, which is contributing to developments in the visual arts.

More than fifty years after the research directed by Bourdieu on the "social uses of photography", the mobile phone seems to be outlining a new multimedia "middle-brow" art. In reality, what sorts of images are produced with mobile phones? What meaning do they hold for their creators and their various "audiences"? In what way do they modify the social interactions that they mediate? Starting in the 1980s, Roland Barthes already considered "pleasure through images" to be one of the defining characteristics of contemporary societies. The present enthusiasm for mobile phone (audio)visual practices reinforces the predominance of the photographic regime, but ultimately, they may paradoxically weaken it. Does the mobile phone not banalize photography by circulating images through various spaces, digital or not, "to the point that it [photography] is no longer faced with any image from which it can be distinguished, relative to which it can affirm its specificity, its scandal, its madness"? (Barthes, 1980, p. 182) [translation by the author].

Mobile images cannot be correctly analyzed without taking their conditions of production into account. Both technically and practically speaking, the telephone combines two devices: it is at once a production tool (a still and/or video camera) and a broadcasting tool (a photo album, a gallery, a portable player, or a medium). Showing a film on a mobile phone has become common enough that it is sometimes considered to be the "fourth screen" (after cinema, television, and the computer).

The Multimedia Messaging Service (MMS) is not the principal use of mobile photography in everyday life, and it seems to be overrepresented in qualitative studies that use quasi-experimental methods (Koskinen, 2005; Koskinen, Kurvinen & Lohtonen, 2002; Ling & Julsrud, 2005; Voida & Mynatt, 2005). Okabe and Ito explain that these studies tend to focus on anticipating future technological developments that are not yet widespread among consumers. They instead employ "naturalistic observation of camera usages" in their ethnographic research (Okabe & Ito, 2006). In this approach, the analysis of "photographic acts" (Dubois, 1990) and image-viewing situations is not separated from the analysis of images themselves, of which a significant corpus has been built through extensive field research conducted over several years' time in a collective framework.[4] The results presented here thus combine the anthropological and semi-pragmatic approaches that were employed.

---

[4]   This work was part of two collective research projects commissioned by the AFOM: Jeanneret & al., and Jarrigeon & Menrath, 2007. Associating direct observation in contrasted, significant situations with in-depth semi-directive interviews, we sought to describe and analyse contemporary uses of the mobile phone, without neglecting the material nature of the object itself and

The mobile phone is far from initiating a break with our motion picture heritage, and, in fact, relies strongly upon it. The intent of this approach – an analysis of the continuities and discontinuities provoked by new visual uses of the telephone – is to derive the aesthetic template common to a large number of mobile productions. Based on studies of various cases including family photos, citizen journalism, everyday life ethnography, spectacular performance, or reflexive use, I shall form hypotheses about the way mobile phone pictures fit into our historical visual culture. This will lead to another crucial question: how does this style of images affect our sense of representation as well as our social interactions?

2   Mobile Images: Amateur Images?

As a "middle-brow art", mobile phone-based multimedia production is essentially practiced by amateurs. There are numerous professional downloadable content creators, but a large part of visual creation and the organization and collection practices characteristic of mobile phone usage are indeed carried out by amateurs. The term is used here in a general sense: an "amateur" is, of course, not a "professional",[5] but is also not an "artist". On this subject, Roger Odin (whose perspective I adapt for my own use) notes that "the professional/amateur opposition is related to the author/spectator opposition and tends to mask the contemporary redefinition of spectator postures" (Odin, 1999, p. 43) [translation by the author]. The mobile phone shifts borders and invites us to question contemporary spectator activity, just as so-called family photography and film, for example, have done before it. A veritable "reactor" (Jarrigeon & Menrath, 2007), it tends to transform the old-style spectator, who is considered to be passive, into a genuine participant in what's happening. Laurence Allard (1999) generally considers the amateur to be a "full-fledged figure in aesthetic modernity", a cre-

---

the audiovisual productions which it begets. The first study dealt with the world built around the mobile phone by its representation in cinema, literature, and advertising. It is in the second study that I concentrated more specifically on mobile visual and audiovisual productions. I analyzed a corpus of mobile films created or presented by artists for different festivals like the Pocket film festival created by the Forum des images (Paris) and Mobile Film Festival (Paris). Available: http://62.210.119.76/ [May, 15, 2011]. Available: http://fr.mobilefilmfestival. com/video.php [May 15, 2010]. I also included mobile films distributed on internet sites like YouTube, Dailymotion, and personal blogs, as well as still or animated images which were shown to me in the course of field research. I did sixty interviews in Paris and other French cities, as well as the countryside, and conducted observations of people in different social situations.

5   This is particularly relevant, for example, when comparing "citizen journalism" carried out with the mobile phone and "full-time journalism".

ative figure that helps us understand the importance of practices catalyzed by the mobile phone. Thus, unlike other approaches that distinguish between artistic and ordinary usage (Koskinen, 2007; Ling & Julsrud, 2005), the intent is not to separate images with "artistic" goals from those that are much more trivial or functional (*i.e.*, the "unpretentious" or even completely "failed" images that are generally produced with mobile phones). I shall consider the full range of visual creation processes while still paying attention to their specific features. New non-telephony mobile phone uses yield a rich repertoire; this continuous spectrum of photographic and video uses includes everything from simple tests to assumed, instrumentalized mastery.

2.1 First Steps and Discoveries

One element in this repertoire is exploratory uses, in which the user tests the potential of the mobile phone with no intent other than to "find out what it can do". Photography and video tests are among the larger set of actions carried out by those who are interested in knowing the technical capabilities of their telephone. For example, a 40-year-old skilled worker named Bertrand explained:

> "I mostly tried taking photos at the beginning, I mean the first day, to see how it worked. Since then, I have to admit that it doesn't interest me much. I didn't take photos before, so..."

Pierre, 15 years old and the happy owner of the "latest Sony Ericsson walkman", is in the same situation. He has made a few short photomontage videos to explore the different "options" on his telephone. The first time we met him was immediately following his purchase, and he was proud to show me what he knew how to do. Three months later, he had completely stopped:

> "I wanted to have it, but then I realized I don't like it. When we need to take photos, it's usually other friends who take them, and they send them to me later – that way I can have them."

2.2 Occasional Uses

Our observations and interviews indicate that a great number of photographic uses are, in fact, occasional, as 25-year-old Nicolas explained:

"Well, I don't use the camera much. Plus, the video is kind of complicated, you have to go through all kinds of settings. In fact, something really has to be going on, or someone has to ask me to use it. But often in those cases, I pass my telephone to someone else."

The mobile phone allows moments to be captured when they are deemed worthy of being photographed. These are most often celebrations or meetings of an exceptional nature, though in the case of "evenings with friends", what is described as exceptional may actually occur rather regularly. For example, Charlie, 17 years old, explained that

"I only take photos with my mobile for my mates, when we have parties you have to take photos. It's for exceptional things. Then again, it's almost every weekend!"

Charlie is among those who clearly distinguish between "old style" photography, which he practices with his father's film camera, and mobile phone photography, which he reserves for marking events connected with friendship and social life. This dichotomy is frequently present in the interviewees' discourse: it corresponds to a practice of mobile phone photography that is marginal but occurs regularly in the organization of everyday life. Mobile phone photography thus follows the rhythm of small communal events.

2.3 Functional Uses

Today, mobile phones give a larger number of people individual access to photography and video. Some, like 30-year-old trumpet player Julien, are quite happy to finally have access to these tools:

"I managed to get a great mobile phone, that way I can finally do what I want. I can even film bits of concerts and practice sessions. Well, it's not always great, but it's practical anyway. What I often do is take photos of sheet music, then I send them by Bluetooth and the whole group can have them!"

Julien has invented a personal functional use, and his mobile phone now participates in the organization of his professional life. He appreciates being able to circulate information and uses his mobile phone as a scanner. Other uses of this type are developing, as in the following scene observed during a winter sale in a women's fashion store: A young girl is admiring a jacket with her friend.

"Do you think I should buy it?"
"It's nice, it's useful to have an everyday jacket like that."
"Yeah, and plus I don't have one. But if I don't ask my mother, she's going to kill me. I have an idea – I'm going to send a photo to my sister, that way she can show Mum and I'll know what she thinks about it."

The two girls then set about photographing the jacket as discreetly as possible. They kept it with them and continued to browse the store while waiting for the mother's response.

Subjects thus find applications for photography when subjected to practical imperatives, and smart phones have amplified this kind of use. It is now quite common to see people photographing a shop's opening hours or wall labels at a museum exhibit. Before the camera phone was omnipresent, they would have most likely noted this information on paper.

2.4 Careers[6] toward Aesthetic Photography and Video

Possessing a mobile phone can also lead users toward more artistic practices. This is, for example, the case of 28-year-old Emmanuel. He had never really done any photography before. Happy to have a small still and video camera in the palm of his hand thanks to his telephone, he went through a short period of technical tests and then set about making portraits, paying careful attention to framing and lighting. He reflected on poses, the composition of images, and the rhythm of surfaces, among other things. One day, I even found him lying on the floor in the stairway of his apartment building. He was trying out different methods to capture the play of light and shadow. Emmanuel's development in the field of photography passed through steps that are in some way classic: from a discovery use, he moved on to systematic exploration, then to the desire to create photographs that take a back seat to the content that they capture, before truly understanding the formal and expressive possibilities of this kind of art.

He was initially disappointed to see that his photos were blurred once transferred to and viewed on the computer, but he then tried to take advantage of these "faults". This transition from the desire for a classic, beautiful image to a more open consideration of what exactly an image is reveals the artistic dimension of these practices. Artists who attempt mobile phone art are generally "experimental" artists, genuinely preoccupied with the particularities of images produced by telephones. Imprecise focus, pixilation, and blurring due to movement of the device are some of the characteristics that remind them of Jean-Luc

---

[6]   Here I use "career" in the interactionist sense of the term (Becker, 1985).

Goddard's famous saying: "Not a just image, but just an image". From this point of view, mobile phone photography and video belong to the category of pictorial or abstract images, and not that of "bad photographs."

The ability to question the device without being disappointed that something is missing relative to existing photographic devices is probably the only factor that clearly distinguishes amateurs from artists who produce work for mobile video festivals or meetings. For example, several art schools have added film projects to their programs that are to be produced with mobile phones. This is the case at the Fine Arts School of Grenoble, L'ECAL (the Cantonal Art School of Lausanne) and the national contemporary art studio Le Fresnoy. Alain Fleischer, its director, is himself the creator of several "pocket films" presented on the big screen during the Forum des Images festival. At its second edition, he explained his choice to make students work with mobile phones as a manner to reflect upon "the adaptation of a tool to a subject".

Though these photographic and video uses differ from the amateur practices that they extend and reinvent, there is nonetheless a certain continuity. In no case is there a logic of substitution: the mobile phone does not replace existing video or still cameras.

3   "Precarious" Mobile Images

Inseparable from their conditions of production and distribution, mobile phone images are obviously marked by certain evolving technical constraints that make them particularly "precarious" images. I have borrowed this expression from an essay on photography by Jean-Marie Schaeffer (Schaeffer, 1989) because mobile phone images are, first and foremost, photographs. As photographs, they have a specific kind of connection to the real, the referent, or a subject. However, mobile phone images are even more "precarious" because they possess a certain fragility (Koskinen, Kurvinen, & Lehtonon, 2002).

3.1 Images of Lesser Value

People consider images made on a mobile phone to be less valuable, which does not necessarily prevent them from being objects of affection.

"You have to admit that they aren't real photos!"
"Anyway, you don't print these images."
"I don't even know where they went, I think I erased them by accident."

All of these expressions tend to discount mobile phone images relative to other photographs. They are indeed rarely printed and are subject to mobile-specific hazards. They are lost, erased by mistake or to make space, and forgotten in the depths of interface menus. Even individuals who keep their mobile photos organized or archived on a computer tend not to ascribe them much importance. This circulation itself, the metamorphosis of passing from one medium to another, contributes to their fragility by desacralizing them as images.[7]

3.2 Honestly, You Can't See Anything

"A mass of pixels", underexposed images, rough framing, uncontrolled movements of the camera... In practice, mobile phone images are far from having the resolution that is constantly claimed in advertisements. Often, one must admit that "you can't see anything" in them (Arrasse, 2003). Bernadette, 50 years old, is amused when she tells us:

> "The other day, a colleague showed me photos of his children on his mobile phone. He was really proud of them... I was a bit embarrassed, I couldn't even tell the little boy from the little girl!"

One of the great paradoxes of amateur image use is the *simultaneous desire for technical perfection* (connected to the promises in advertising) and the *denial of inferior quality* in reality. Bernadette's colleague was quite proud to have a telephone with "I don't even know how many pixels", but did not see that his photos were of very low quality.

Let us now look at the case of Laura, a 14-year-old adolescent who really "pestered" her parents to get her a mobile phone with a "good camera" for her birthday because she was going to a Bénabar concert and is a big fan of his. Her persistence paid off, and she went to the concert well equipped. After vaunting the merits of her mobile phone's camera, she showed me her images, in which one distinguishes a vague silhouette on a stage saturated with light. "Wait, I'll show you more detail!" She began to zoom in, and the face of what was supposedly Bénabar blurred increasingly as it enlarged to fill the screen.

In practice, the denial of low image quality allows happiness about having been able to make the image as well as initiating a discussion of the concert experience using the photo as a prompt. In many cases, *the photographic act supersedes the photograph itself*. The image triggers comments that are initially

---

[7]   This explains the economic failure of the Orange/Photoservice partnership. Relatively few clients came to print their mobile phone photos at these photo development stations.

a description of what should be seen. In turn, and more importantly, this serves to *initiate remembrance* and *engage conversation* about a personal experience.

One of the characteristics of silver photography brought to light by Roland Barthes – the index point – is accentuated in mobile photography (Barthes, 1980). The "that has been" Barthes describes is perhaps the principal mechanism behind these mobile practices. Mobile images are, in fact, destined to be shown (preferably on the mobile phone itself) and allow Laura to discuss an event and tell her story using her low-quality photos of Bénabar.

4   The Everyday Production of Special Events

The mobile phone stands out among photographic capture and display tools. It cannot be considered equivalent to other devices because few objects, technological or not, are so continuously situated within the bodily sphere of their users (Jarrigeon & Menrath, 2007). This omnipresence increases the number of opportunities for photography, as it is the only photographic device that allows its user to capture images without planning.

4.1 Creating Photographic Opportunities

Unexpected use falls under the "kairos regime", which is to say the regime of chance and opportunity (de Certeau, 1990). That is, the unplanned and the surprising are likely triggers for photographic acts. In this way, the mobile phone acts as the enabling factor for a shared fantasy wherein we imagine ourselves to be "making films all the time", to borrow an expression from artist Christophe Atabékian.[8]

In a way, the mobile makes expression of the "autobiographical impulse" possible (Cooley, 2005). It can potentially transform one's life into a series of rehearsals and takes, a studio in which images are collected for an upcoming album or film. However, in practice, one wonders what should be done with these images once they have been shown in raw form on the mobile's screen. Among the people I interviewed, there was no lack of plans to archive them, but the work was often put off for later. The musical group Mogwai humorously reinvents the real through collage, combining different moments captured from

---

[8]   Christophe Atabékian, during the 2nd edition of the Pocket Film festival. He also proposes "artistic performances" in which he creates one film per day, every day, using mobiles that are lent to him. Available: http://lesfilmsdepoche.com/pocketfilms/category/christophe-atabekian/[May 15, 2010].

life. For example, they have made a fictional concert on mobile phone from different concerts recorded over the course of several years.[9]

4.2 Ordinary Testimony or Amateur Journalism

This multitude of photographic opportunities tends to transform mobile phone users into small-time journalists (Okabe & Ito, 2006) and ethnologists of everyday life. Once equipped, they lie in wait for something that could become a notable event.

The mobile is often considered to be a new form of "pen-camera" that facilitates *improvised reporting.* Extraordinary events recognized or lived as such are favored moments for using mobile phones. In France, the social movements of 2005 and the extensive demonstrations against the First Employment Contract (*contrat première embauche*) were often captured in images by participants who then posted them on official newspaper blogs to reach a wider audience. The confrontations between police and individuals at Paris' Gare du Nord station at the beginning of 2007 also yielded a large number of mobile phone images that were subsequently posted on various web sites. Some of these images were problematic due to the violence they depicted and were therefore rapidly deleted. Others circulated widely, particularly via *Youtube* and *Dailymotion.* As yet another example, as I am finishing this article, it is snowing over all of northern France, and the press is asking for amateur images to illustrate the hourly stories about blocked roads.

The mobile phone is thus an essential element in the current redefinition of the border between professional journalism and what supporters call "citizen journalism". However, a law concerning the "prevention of delinquency" enacted in April 2007 halted this movement and may represent a step backward.[10] It includes an amendment addressing the subject of "happy slapping", *i.e.,* assaults recorded and broadcasted in almost real time. The amendment explicitly prohibits the recording of violent images, including those that could incriminate representatives of the State. This restriction does not concern professional

---

[9] Available: http//www.mogwai.co.uk [May 15, 2010].

[10] The text sets penalties of up to five years in prison and a 75,000 euro fine for distributing images depicting the infractions mentioned in articles 222-1 to 222-14-1 and 222-23 to 222-31 of the penal code. These infractions include everything from extreme violence ("torture" and "acts of barbarism") to simple aggression. The article 222-13 concerns violence "committed by representatives of public authority (...) in the exercise (...) of their functions". The law specifies that this prohibition "is not applicable if the recording or distribution is a result of the normal exercise of a profession whose objective is to inform the public, or is carried out in order to serve as proof in court."

journalists and could thus rapidly halt the general movement towards greater public participation in media production.

We must recall that amateur testimony covering important events is not at all a new phenomenon; the best-known images of trench warfare during the First World War were made by anonymous individuals. The question of a photographer's complicity that is captured by the camera is indeed an integral part of the history of documentary photography.

4.3 Small Images and Short Formats: Optical Haiku

Other relationships to events exist beyond these quasi-journalistic practices. The mobile phone often serves to *transfigure the banal* and the everyday into something outstanding. It gives rise to a greater attention to the *picturesque* (*i.e.*, to that which is worthy of being painted). For example, Martine, a 35-year-old independent country nurse, tells us:

> "I was in the car, it was cold and the whole landscape was frozen. I saw a tree that surprised me. I wasn't in a hurry, so I stopped to take a photo. It's funny, I would never have done that before!"

In many situations and in many different ways, mobile productions take the form of *amateur haiku*. Like this classical Japanese art, the mobile phone permits an "art of very little" (Keblaner, 1983), both because it allows minor situations lived by the individual to become premises and because it produces short, condensed forms. Like the *haiku*, it allows us to *transfigure the everyday through calculation and spontaneity*, although it obviously uses more trivial means. This dimension can also be found in older uses of the mobile phone such as SMS. As an abbreviated form, it requires special attention to the relationships between observations, lived situations, and a message that is sometimes destined to be kept by the recipient.

"Calm sea. Champagne. Happiness", Martine writes to her friends. She is in Biarritz, where she has gone on an amorous escapade with her boyfriend. Anne receives a supportive message: "Bad weather. Hang in there, just a few meters to go."

5   The Desacralization of Family Photographs and Films

The growing multitude of opportunities to use photography and video and the instantaneous character of mobile practices are playing an increasing role in how we construct images of personal life, and in particular family life.[11] In the 1960s, Pierre Bourdieu and the research team that he coordinated (1965) demonstrated that the primary social function of family photographs was to consolidate or even produce the myth of the family. Today, the mobile phone plays a part in the ongoing desacralization of these practices and extends the "Kodak culture" effects analyzed by Chalfen (1987) in the eighties.

In the 1960s, the myth of the family was primarily an origin myth: with group photographs, one was able to produce evidence of heredity through the interplay of resemblances and differences. Since then, family uses have opened up to photographic moments that are somewhat less official than births, baptisms, and marriages. An increasing variety of moments are considered worthy of being photographed: not only the "big" events, but also other festive times such as birthdays, Christmas, or vacations. The birth of a child is also an important motive for acquiring photographic or video equipment.

The emergence of digital photography has brought great change by permitting a larger number of photographs to be taken. However, the occasions for digital photography generally remain conventional and repetitive, and one usually does not get the camera out without a good reason. Events are photographed when they are considered to be key moments providing evidence of familial happiness. Family photographs are like home videos in the sense they show very little intimacy; crises, tensions, and disputes are eliminated in favor of eternal smiles, light-hearted poses, and programmed embraces. From this point of view, the mobile phone seems be opening a new phase in the movement toward desacralization. More spontaneous (at least for those who use them!) and more apt to capture moments humorously, they allow a new sort of family image to be produced, in which ridicule and the grotesque, for example, have an important place. Consider this observation made in a family setting:

> "At the crucial moment where the youngest member of the B. family is going to blow out the candles, everyone takes out their camera. But those who instead 'draw' their mobile phones do not seem preoccupied by the idea of creating the same sort of images. Cousin Pierre, for example, amuses himself by taking photos of the others,

---

[11]   One should not forget that photographs and films of families were present from the very beginning of these media. Photographers made portraits of those around them. Jacques-Henri Lartigues, today considered to be one of the great artists of the beginning of the 20th century, was an amateur, and his subjects included his friends and family in their daily lives. Among the first films made by the Lumière brothers was "Baby's lunch!"

who are themselves taking photos. He immediately shows these images to his sister and both laugh, saying: 'Ah, did you see Uncle Bernard, do you see the face he's making?'"

Apparently, a space is being created for more open and spontaneous family images and for a plurality of family stories to come into play.

6   From Staging to Performance: The Inner Workings of Mobile Sensationalism

Though it makes the immediate production of more spontaneous images possible and contributes to the desacralization of familial photographic practices, the mobile phone does not prohibit staged action (*mise en scène*). Mobile phone home videos provide particularly convincing evidence of this. The absence of narrative construction is characteristic of family films (Odin, 1999); in mobile phone video, overlong passages disappear and narrative emerges, as amateur filmmakers are forced to adapt to small, short formats. These short films are much denser in terms of action than those produced with a video camera, which are exciting to make, but much less so to watch (most of us have had the experience of being obliged to watch our uncle's endless cassettes at one time or another). Within the space of a few seconds or a few minutes, mobile videos present scenes that are more structured, more focused on a single moment, or that lead directly to a punch line. A beginning, a middle, and an end: according to Aristotle's ancient precepts, these are the ingredients of drama, and they are being rediscovered (perhaps involuntarily) by users of mobile phone video functions.
    These short films are thus clearly distinguished from classic films by their improvised staging, which is related to an often intuitive grasp of the format. On the other hand, they retain a certain visual common ground including motion blur, overly rapid panning, and acute "zoom-itis" when the equipment allows it. The temporal laxity and blur characteristic of family films allow each person to produce his or her own interpretation of the family. Roger Odin clearly demonstrates how these "defects" are, in fact, intrinsic necessities of the family film, as they allow those who do the recording to avoid imposing a certain meaning of familial relations upon other members (Odin, 2004). Newer, mobile video productions also require reflexive (internal) conflict to be 'defused'. This necessity is not so much a result of the films themselves as how they are produced and shown: right away and without planning. In the past, one would set up equipment and even a sort of scenery for family film sessions. Mobile films, though

more structured, allow a plurality of sometimes contradicttory points of view –
notably the emergence of female viewpoints – whereas the video camera often
remained in the hands of the father. In this way, according to Roger Odin (1999),
"they are perhaps better objects for the family."[12]

6.1 Recycling Media Referents

If mobile phone *mise en scène* is starting to stir things up in the domain of home
movies, it is already essential in cinematographic practices among friends, where
media referents from diverse sources intervene. The most obvious example is
probably the universe of advertising. It greatly inspires youths, particularly adolescents, who appropriate it to humorous effect. For example, Marie and her
friend Géraldine, 15- and 16-year-olds, respectively, amused themselves by re-
making various shampoo commercials, whereas Pierre and Julien, of the same
age, mimic the "sports action" around which Nike or Adidas advertisements are
centered. Both pairs of friends sing their own interpretations of slogans, modifying the words and mimicking gestures. The scenery receives less attention: a
simple exterior shot for the boys, who employ a parcel of grass, or "interior –
bathroom" for the girls, who are in any case filming themselves close-up.

Additionally, many videos reproduce the codes of certain television series
or news programs. Long shots, off-camera voice commentary, and intonations
indicate how much these young people are amused by actively re-appropriating
images produced in more institutional frameworks.

6.2 Happy Slapping: Morbid Mise en Scène between Gag Videos, Snuff Movies, and Jackass

Staging is an important element when practices are in some way performances.
Sensationalism is a particularly effective mechanism and can take a multitude of
forms, from the banal to the extremely violent. The mobile phone allows the
combination of staged action with its instantaneous execution. It is a tool that can
incite people to do the audacious, as in this typical scene overheard in the metro:

Two boys are commenting on a video on a mobile phone that they are passing back and forth. One says to the other:

---

[12]   Jarrigeon and Menrath (org), « Le téléphone mobile, une affaire de famille? », Round table
         discussion organized for the AFOM at the *Maison de la Chimie*, Paris, 3 April 2009. cf Mobile
         et société no. 8, July 2009. Available: http://www.afom.fr/eclairages/famille-et-telephone-
         mobile [September 15, 2010].

"You'll see, I dared to do it, in the middle of the store, my boss was right there, but oh well, I did it anyway."
"No way, you turned the music up that loud?"
"Yeah, and look, I'm dancing like crazy, it was so funny, like a music video in the middle of the store!!! It was my other colleague who recorded it, he was dying laughing. Ah, see there, you hear it..."

In this case, the video seemed innocent enough, but staged action can sometimes border on pathology or seriously threaten one's own safety and that of others. There are numerous "pocket films" that make use of a *register combining the grotesque and the macabre.* This register is not new, and the media have been exploiting this brand of humor particular to amateur films for quite a while, notably in the "funniest home videos" series that chains together images of falling babies, domestic accidents, fainting brides, and spectacular sporting mishaps.

This fusion of the grotesque, the playful, and the macabre is at the heart of Johnny Knoxville's business. In 1997 in the United States, he invented the "Jackass phenomenon" by devising and recording risky and disgusting exploits, such as spraying pepper into his eyes, sitting without pants on the stinger of a scorpion, jumping off the roof of a house on a bicycle, or diving into a swimming pool full of excrement. In 2002, MTV began broadcasting his show, which had considerable success not only among stunt lovers like skaters, but also among a more general audience.[13] Certain practices spread through the population and fed into this particular aesthetic, which is especially likely to be seen on the screens of mobile phones.

During our investigation, we had the occasion to view a great number of these kinds of images on the mobile phones of our interviewees. Even when images did not directly evoke this referential, viewer commentary may refer to it, as in the following example:

During an evening among friends, 32-year-old Fabrice approaches us with his mobile phone.
"Wait, take a look, I have something great to show you. I have a real snuff movie."
We all approach him, a bit surprised, because this is not really something he would be into – unlike Antoine, who the other day had shown us scenes of fights, which were apparently real and filmed in the entry hall of his estate in Cergy Pontoise. He opens the film. It is a distorted close-up of the face of a friend, that we have difficulty recognizing.

---

[13]   We recall that extreme sports enthusiasts have long delighted in sensational images of this sort, and that skate videos always include a chapter showing the most impressive failures and the most spectacular accidents.

"Check out this snuff movie! Karim, recorded live, with no special effects... sleep-
ing!"

This scene reveals the banality of mobile-phone humor's morbid clockwork,
which is not entirely alien to the phenomenon of happy slapping. This type of
humor conforms to an aesthetic template, which pre-exists yet greatly transcends
it. This can perhaps explain the relative success of happy slapping images,
which, as our survey of young people shows, circulate rather freely from one
mobile phone to another, in parallel with equally violent images of current
events, such as those at Abu Ghraib or the recorded execution of Saddam Hus-
sein.

Young people in particular sometimes watch these images repeatedly and
look to them for inspiration when creating their own imitative scenes. After
showing us the video of a teacher being attacked in Porcheville (which was ex-
tensively covered in the French press), Sylvain, 15 years old, explained to us that

"Once, one of my friends did something great. She pretended to get slapped, and had
someone make a video of her falling on the ground. It was fake, but really well
done. I don't know how she did it."

Gilberto, 14 years old, tells us that

"I know people who act like they've had an accident at the side of the road, for fun.
They make videos of how people react when they stop, and count the cars that don't
stop. It's just a game, but well, something bad could end up happening!"

There is indeed an *aesthetic template* and, thus, a certain continuity between
simple staged action, playful imitation, and happy slapping videos in the strict
sense. However, we should not be too hasty to declare the banalization of evil,
nor too directly incriminate the tool itself. Though the mobile phone assists in
the production and circulation of hard-hitting images, we must take care to
clearly distinguish the different levels of involvement that social subjects can
have in these uses. The psychiatrist and psychoanalyst Serge Tisseron has shown
that it is essential to take the *viewing conditions* of violent images seriously in
order to understand what is at stake in our relationship with them, and more
specifically, what is at stake for children and adolescents (Tisseron, 2003). The
act of collecting, showing, or even drawing inspiration from violent imagery for
fictional or playful imitations does not imply approval, but rather a form of *distance, which operates through manipulation and action.* With the arrival of the
mobile phone among existing aesthetics, the question of image literacy education
must be re-examined.

7   A Tool for Reflexivity

The mobile phone invites us to question our relationship with images and the devices that produce them. It is a tool *for reflexivity* on several levels. First, it allows us to establish new forms of introspection: family and friends are favored subjects for mobile phone images. *Self-portraits* also have an important place, both in uses with no real artistic intent and in those that are openly aesthetic. We photograph and film ourselves in close-up, we watch ourselves look into the curious lens, which points to another reference: the *photo booth*, in which we squeeze together, close to friends and family, cheek to cheek, to fit everyone into the frame.

On another level, it is also often the telephone itself and its uses that inspire mobile productions. This is one of the recurring themes for those who present their films at festivals: they use the mobile phone to analyze mobile practices. It would be possible to give many examples that in a way reveal contemporary mobile uses *ad absurdum* (Jarrigeon & Menrath, 2008, p. 192). However, we shall only mention two of them here. The "versatility of the object" is quite often represented in a humorous way, as in the film *Coup de fil rasoir* ("Dull phone call") by Ronan Fournier-Christol,[14] wherein the director shows himself shaving in front of a mirror with his telephone in hand. His phone rings, and he responds that he can't hear anything, he's late, he's shaving, and that his correspondent should speak louder. Another film presented on YouTube shows the multiple uses of the mobile phone, serving as both a razor for the husband and a steam iron for the grandmother. The film ends with an exterior scene where we see the grandmother taking a walk with her telephone in hand, and what looks vaguely like an earphone hanging from her neck. A thief on a scooter snatches her mobile phone; she presses the earphone and the scooter explodes.

As another example, way-finding practices and the fabrication of memories in the present are at the heart of Alain Fleischer's Chinese Tracks,[15] which wanders for half an hour through a Shanghai neighborhood undergoing reconstruction. The narrator has been in the neighborhood the night before with a friend who lives there, and he hopes to find a grey door behind which, he says, "is the most beautiful girl in Shanghai". Taking seriously the fact that a telephone is not an ordinary camera, but a device "without a viewfinder, and with no real framing", Alain Fleischer asks himself what "filming with a telephone means – is it not basically 'filming with the ear'?" He progressively loses himself in the narrow neighborhood streets and comments on his wanderings as though he is being

---

[14]   Fournier-Christol R., *Coup de fil rasoir*, 2005. Available: http://62.210.119.76/spip.php?article155 [May 15, 2011].

[15]   Fleischer A., *Chinese Tracks*, 2006.

guided from afar. In doing so, he continuously captures video with the telephone held at ear-level. He stops after half an hour of continuous performance, finding himself in front of the door in question. He describes it, but on the screen we observe only the wall that the ear would see, if it could see.

8 Conclusion

Artistic productions often focus on the technical and social specificities of the mobile phone, and establish this device as a contemporary *aesthetic experience analyzer*. They ask us to take seriously the effects of the mobile on pre-existing visual production methods, which continue to be its contemporaries. The practices analyzed here dialogue with other, older ones, shifting the "social uses of photography" (Bourdieu et al., 1965). This new *middle-brow multimedia art* contributes, as we have shown, to a *reconfiguration of contemporary spectator activity*, shifting our relationship with images bit by bit, but also more generally modifying our relationship with our perceptible environment, visual or auditory.

Photographable situations and the various moments of audiovisual sharing, during which a group of people crowd together over a single telephone, transform a supposedly individual object into an instrument of conviviality. These situations often take precedent over the images themselves. From this point of view, the (audio)visual uses of the mobile telephone have a place among mobile phone sociabilities, which extend well beyond the communications framework for which it was designed and in which it is most often studied.

References

Allard, L. (1999). L'amateur: une figure de la modernité esthétique. In R. Odin (Ed.), *Communications n°68: Le cinéma amateur* (pp. 9-33). Paris: Seuil.
Arrasse, D. (2000). *On n'y voit rien. Descriptions*. Paris: Denoël.
Barthes, R. (1980). *La chambre claire. Note sur la photographie*. Paris: Seuil. [Barthes, R. (1981). *Camera Lucida. Reflections on Photography*. New York: Hill and Wang].
Becker, H. (1963). *Outsiders. Studies in the Sociology of Deviance*. New York: The Free Press.
Bourdieu, P., Boltanski, L., Castel, L., Chamboredon, J.-C. (1965). *Un art moyen. Essai sur les usages sociaux de la photographie*. Paris: Ed. de Minuit. [Bourdieu, P. (1990). *Photography: A Middle-brow Art*. Stanford: Stanford University Press].
Chalfen, R. (1987). *Snapshot Versions of Life*. Bowling Green: Bowling Green State University Press.
Certeau, M. de (1984). *The Practice of Everyday Life*. Berkeley, CA: University of California Press. (Original work published 1980)

Cooley, H. R. (2005). The Autobiographical Impulse and Mobile Imaging: Toward a Theory of Autobiometry. Paper presented at the Workshop *Pervasive Image Capture and Sharing: New Social Practices and Implications for Technology* at Ubicomp'05 Shinagawa, Tokyo September 20. Available: http://ubicomp.org/ubicomp2005 [May 15, 2008].

Dubois, P. (1990). *L'acte photographique*. Paris: Denoël.

Jarrigeon, A., & Menrath, J. (2007). *Le téléphone mobile aujourd'hui: usages et comportements sociaux*. Research paper. Available: http//www. afom.fr/v4/STATIC/ documents/rapport_gripic_integrale.pdf [May 15, 2008].

Jarrigeon, A., & Menrath, J. (2008). La part du possible dans l'usage. Le cas du téléphone portable. *Hermès, 50*(2), 99-105.

Jeanneret, Y., Menrath J., & Lallement E. (Ed.). (2004). *Le téléphone mobile aujourd'hui, usages, représentations et comportements sociaux*. Rapport de recherche. Available: http://www.afom.fr/v4/STATIC/documents/RAPPORT_GRIPIC_2005.pdf [December 15, 2005].

Klebaner, D. (1983). *L'art du peu*. Paris: Gallimard.

Koskinen I. (2005). Seeing with Mobile Image. In K. Nyíri (Ed.), *A Sense of Place* (pp. 339-348). Vienna: Passagen Verlag,

Koskinen, I. (2007). *Mobile multimedia in action*. New Brunswick: Transaction Publisher.

Koskinen, I., Kurvinen, E., & Lehtonon, T. (2002). *Mobile Image*. Helsinki, Edita: I T. Press.

Licoppe, C., & Zouinar, M. (Eds) (2009). Les usages avancés du telephone mobile. *Réseaux, 27*(156).

Ling, R., & Julsrud, T. (2005). The Development of Grounded Genres in Multimedia Messaging Systems among Mobile Professionals. In K. Nyíri (Ed.), *A Sense of Place* (pp. 329-338). Vienna: Passagen Verlag,

Odin, R. (Ed.) (1999). *Le cinéma amateur. Communications n°68*. Paris : Seuil.

Odin, R. (2004). Les films de famille: de "merveilleux documents"? Approche sémiopragmatique. *Actes du Colloque Le film de famille*. Bruxelles: Publications des Facultés universitaires Saint-Louis.

Okabe, D. & Ito, M. (2006). Everyday Contexts of Camera Phone Use: Steps Toward Technosocial Ethnographic Frameworks. In J. Höflich & M. Hartmann (Eds.), *Mobile Communication in Everyday Life: Ethnographic View* (pp. 97-102). Berlin: Frank & Timme.

Okabe, D., Ito, M. & Matsuda, M. (2006). *Personal, Portable, Pedestrian. Mobile Phones in Japanese Life*, Cambridge, MA: MIT Press.

Rivière, C.-A. (2005). Mobile Camera Phones: A New Form of Being Together in Daily Interpersonal Communication, in R. Ling & P. Pedersen (Eds), *Mobile Communications: Renegociation of the Social Sphere*. (pp. 167-185). London: Springer.

Rivière, C.-A. (2006). Téléphonie mobile et photographie: Les nouvelles formes de sociabilités visuelles au quotidien. *Sociétés, 1*(91), 119-134.

Schaeffer, J.-M. (1989). *L'image précaire: Du dispositif photographique*. Paris: Seuil.

Tisseron, S. (2003). *Enfants sous influence: Les écrans rendent-ils les jeunes violents ?* Paris: Armand Colin.

Voida, A. & Mynatt, E. D. (2005). Six themes of the communicative appropriation of photographic images. *Proceedings of the ACM Conference on Human Factors in Computing Systems*. New York: ACM Press, 171-180.

Visual Mobile Phone Content and Developmental Challenges

The Mediatization of Social Relationships in Adolescence

Iren Schulz

1 Introduction

In public debates and scientific discussions, young people are often described as "Generation Mobile" (Schuh, 2007) or as "Digital Natives" (Tapscott, 2009). Not only are adolescents provided with a label, but they are also provided with the "media society" in which to grow up. These labels include two main generalizations. First, they reflect the adult outsider perspective on young people and refer to the integration of youth in societal contexts. Second, these labels emphasize the increasing importance of media for growing up today. In fact, adolescents *are* "digital natives" because they deal with a comprehensive and complex media ensemble including PCs and laptops with internet access, MP3 players and portable PlayStations and, of course, mobile phones (Kraut, Brynin, & Kiesler, 2006; MPFS, 2009). Nearly every adolescent owns at least one mobile phone with a variety of multimedia-based functions and digital services. It is assumed that girls and boys use media to deal with developmental challenges that are very important in this phase of life, especially to negotiate relationships and develop identity, but also to organize everyday life in school and leisure and delineate norms and values.

Against this background, this article addresses the increasing diffusion of new media technologies as a main characteristic of communication in the 21st century and analyzes its meaning for socialization processes in adolescence. The aim is to explore strategies of participation in adolescent relationships by focusing on topics, types and contexts of communication using visual mobile phone content. The theoretical foundation integrates assumptions from developmental theory and phenomenological network concepts and refers to the concept of "Mediatization" as a meta-theory to describe social changes in a mediatized society. Within this framework, adolescent relationships will be conceptualized as dynamic social networks of negotiated meanings. These negotiation processes are contextualized by developmental issues that are typical for this age. Furthermore, mobile phone communication will be examined in line with face-to-face communication and communication carried out using other media technologies, such as television or the internet. The empirical findings support and substantiate the theoretical framework. The results are based on 400 pictures, animations, photos and video clips that are examined together with in-depth interviews with thirteen adolescents age 12 to 17. The final part of the article draws several conclusions concerning the consequences that materialized in the participation processes with mobile content.

2   Mediatization of Communication Processes and Social Relationships in Adolescence: Theoretical Framework

With regard to the comprehensive appropriation of digital media devices and media content in all contexts of everyday life, it is assumed that media-based communication amalgamates more and more with communication practices that were historically unrelated to media. These changes result in new and diversified patterns of communication and have an important influence on socialization processes in adolescence, especially on the negotiation of social relationships.

2.1   Mediatization as a Social Change

The concept of "Mediatization" describes the integration of new media with old media into social life on a meta-level and as a long term process:

> "[...] People think and live their everyday lives ever more connected to these media and the possibilities that are opened up. The concept of mediatization of course includes that this have consequences for people, their everyday lives, their identity and their social relation, as well as for culture, democracy, the economy and society in general because all areas of human life are involved, leisure and work, learning and entertainment, social relations and forms of communication, etc." (Krotz, 2005a, p. 450).

The starting point of these complex processes are the technological changes based on the digitalization and convergence of media. Notably, the mobile phone has evolved into a multifunctional medium that integrates "old media" like radio, TV or print and works as a telephone, game console or as a player for music and video files. Furthermore, it can be connected to a PC or the internet to upload and download media content and to use online services. Modern phone's multiple functions are the reason they seem to not only be a mobile telephone but a portable computer with a connection to the telecommunications network. They provide complex possibilities for communication, entertainment and information based on language, text and images (Krotz & Schulz, 2006). Therefore, it is necessary to find an adequate description of communication using multifunctional, digital media like the mobile phone. Here, Friedrich Krotz refers to the assumptions of symbolic interactionism and describes communication using media as a kind of symbolic action and a kind of interaction:

> "If one understands communication with media as a modification of face-to-face-communication it can be generally asserted that communication with media is also embedded in situations and individuals are also acting from the perspective of a specific role. Every process of communication with media and of understanding is based on an anticipated adoption of perspectives and is shaped through an inner dialogue. In that way, communication with media can be accomplished and understood (Krotz, 2001, p. 74).

In line with that definition, Krotz specifies three types of media-based communication that are also relevant for analyzing communication practices with mobile phones. The first type refers to media-based interpersonal communication, such as by landline phone or e-mail, whereas the second type involves the communication of a person using standardized media content, such as the production of media, for instance, when a person writes a book or makes a film, as well as to the reception of media, as when one reads a book or watches a film. A third type he names is communication with interactive media content, for instance, communication with intelligent software like computer games or Tamagotchies (Krotz, 2007, 2008) In addition, the article takes into consideration types of mediatized communication that are directed at or related to interpersonal relationships, such as when a person carries a mobile phone to show it to everybody or uses that medium as a diary.

These types of mobile phone-based communication have become a part of the social practices of everyday life (de Certeau, 1988). On one hand, communication practices with digital media result in processes of integration on a temporal and spatial level. Permanently available media devices and content are present in all contexts of everyday life and connect different areas of work, leisure or social relationships. On the other hand, mediatized communication refers to processes of social disintegration because the use of digital media like the mobile phone constitutes new motives, purposes and situations which lead to modified and new social arrangements. All of these processes of integration and disintegration represent comprehensive forms and practices of mediatized communication (Krotz, 2001).

The mediatization of social practices is especially relevant for socialization and for developmental negotiation processes in adolescence. The literature gives two reasons for this fact. First, adolescents are known as "Early Adopters" who face new technologies and their possibilities in a very curious and open-minded way. Adolescents undertake multifaceted and creative forms of appropriating media, especially the mobile phone, into their everyday life. The media use of young people also points to technological developments and communicative practices that could be of importance in our future society. Second, adolescents turn toward media to deal with developmental challenges. These processes are increasingly intertwined with mediatized communication and reveal new and changing ways to socialize during this important phase of life. Looking at the communicative practices of today's adolescents gives us an idea of how adults and future generations will communicate in a social context. Against this background, the next chapter provides a concept of social relationships in adolescence that takes the developmental perspective into account and refers to the communicative processes of constituting relationships as social networks. Furthermore, the importance of media, and especially the mobile phone, in adolescent relationships will be analyzed.

2.2 Social Relationships and Digital Media in Adolescence

The time of adolescence is characterized by a simultaneous curiosity toward new experiences and doubts concerning the perception of self in relationships. Young people must confront physical and cognitive changes, especially concerning social issues (Havighurst, 1972). One of the most important social challenges lies in obtaining and differentiating new relationships. Peer-relations are considered the frame of action in which adolescents deal with developmental issues such as forming a gender-specific identity or achieving a certain kind of appreciation. Within peer-relations, girls and boys are able to test different roles and to break boundaries they accepted during childhood. (Oerter & Dreher, 1998) Peer-relations are mostly same-sex ties between adolescents of around the same age that are contracted voluntarily and shaped through collective activities and communicative practices, such as routines to demonstrate status, attempting to stand out from others or renegotiating levels of membership:

> "Everyday activities in preadolescent and adolescent culture enable peers to negotiate and explore a wide range of norms regarding personal appearance and the presentation of self, friendship processes, heterosexual relations, and personal aspirations and achievement." (Corsaro & Eder, 1990, p. 215)

Within and in addition to peer-relations, there are a number of more important relationships in adolescence that are negotiated through communicative practices and interactions. Best friendships are a special type of peer-relation and are characterized by a high level of acceptance, understanding, self-disclosure and mutual advice (Corsaro & Eder, 1990). The status of "always being there for each other" is expressed in shared activities and experiences as well as in support in solving problems and uncertainty (Kolip, 1993). In contrast to best friendship, the first romantic relations are more a "clumsy experiment":

> "In these beginning relationships, the focus is not on the nature of the relationship or the fulfillment of various needs, but on who the partner is, the partner's attractiveness, how they should interact in a romantic context and what their peers think of the relationship" (Brown, as cited in Furman & Simon, 1998, p. 734).

According to these intentions, partnerships do not last longer than several days or weeks and are constituted by playful interactions like dating and kissing but also by communicative practices like disclosing doubts and anxiety (Barthelmes & Sander, 1997; Lenz, 1989). Besides peer-relations, best friends and romantic partners, the re-negotiation of the relationship with parents plays an important role in adolescence. Young people try to attain more independence and autonomy, but their mother and father play an essential part in financial and emotional support. The relationship of an adolescent with his or her parents is characterized by an ambivalent mixture of psychological dissociation and functional attachment (Gille, Sardei-Biermann, Gaiser, & de Rijke, 2006; Shell Deutschland Holding, 2006). Furthermore, young people relate to siblings and other relatives but also to other adults like teachers or coaches (Zinnecker, Behnken, Maschke, & Stecher, 2002).

These relationships are differentiated in adolescence and are negotiated through specific interactions. To integrate these diverse relations and their communicative practices in a theoretical framework, phenomenological network concepts are very useful when taking into consideration the assumptions of symbolic interactionism (Blumer, 1969). For instance, Fine and Kleinman (1983) conceptualize social networks as dynamic sets of relationships and assume that the negotiation of meanings provides the basis for social relationships. The social structure of such a network is constituted through meaningful self-other interactions and communicative negotiation processes:

> "Since meanings provide the basis for individual and collective actions, people's meanings will have consequences for their actions, the production of social structures, and changes within those structures. [...] Understanding actor's meaning is fundamental to any analysis of social structure" (Fine & Kleinman, 1983, p. 98).

Using these theoretical assumptions, adolescent relationships can be conceptualized as dynamic social networks, including different and changing relationships with peers, best friends, partners, parents and other adults. These relationships are negotiated through communicative practices that are increasingly intertwined in processes of mediatized communication.

Historically, media like books, music cassettes, TV series or computer games play an important role for adolescents in negotiating their different relationships. For instance, adolescents use media with their peers and best friends or talk about media. They share media content as a sign of affection for potential partners and use media as an object to demonstrate their inclusion in special peer cultures and create their own styles and preferences that are different than the interests of their parents (Barthelmes & Sander, 1997; Livingstone, 2002; Suoninen, 2001).

With the digitalization of media, more possibilities were created for mediatized communication within adolescent networks. Interactive TV shows, online games, social network sites and instant messenger programs broaden and change communicative practices and therefore the structure and meaning of relationships during adolescence. The mobile phone plays an especially important role as a personal, convergent and multifunctional medium within these relational communication processes. Within peer-relations, girls and boys use the mobile phone to coordinate and sustain these relationships. Furthermore, the phone acts as a lifestyle object for self-presentation among peers (Castells, Fernández-Ardèvol, Qiu, & Sey, 2007; Haddon, 2004; Ito et al., 2008). To socialize with potential partners, girls and boys establish communicative practices that regulate the balance between intimacy and distance. Ling (2004, 2007) describes the mobile phone as a "quasi-illegal medium" that allows adolescents to communicate behind the backs of their parents and provides them with the possibility to develop qualifications for future partnerships. For example, short messages are used to initiate romantic relations and to flirt in secrecy, or these messages work as symbolic gifts to demonstrate sympathy and belonging (Taylor & Harper, 2003). The mobile phone can also be described as an "extended umbilical cord." Mothers and fathers use this medium to retain control and a sense of security when their children go out with friends, peers and partners (Feldhaus, 2004). To communicate with their parents, adolescents mainly use the telephone function of their mobile phone. Communicative practices with peers, friends and (potential) partners include all multifunctional possibilities: sending and receiving text messages; producing, saving and sending pictures, music files and video clips; and selecting logos and ringtones to personalize the mobile phone. In summary, different mobile phone communicative practices have become a constitutive part of negotiating relationships in adolescence.

3 Visual Mobile Phone Content and the Negotiation of Relationships in Adolescence: Empirical Results

The following chapter describes adolescent communicative practices using visual mobile phone content. The findings are part of a longitudinal research project with a qualitative, ethnographical, multi-method design. From 2006 to 2008, three groups of best friends and their social networks were explored. During single and group interviews, observations, media diaries and content analysis, the focus was on the everyday life of these young people and their current developmental challenges, especially regarding relationships, and the mediatized communicative practices they use to deal with these challenges. The empirical data on visual mobile phone content presented in the next chapter is part of this longitudinal study and was conducted in February and March 2007.

In multiple group interviews, the friendship groups (three girls aged 16 with higher-level education[1], four girls aged 13 to 15 with high education, and two boys aged 16 with lower-level education) were asked to show and comment on the visual content they collected on their mobile phones during the last months and years. Furthermore, they were asked to send all of the pictures, animations and video clips via Infrared or Bluetooth[2] connection to the interviewer's laptop. The two girls (one from each girl group) in possession of a mobile phone without the necessary features to receive, watch and send visual mobile phone content were also included in the group interviews to further explore the function and importance of visual mobile phone content for the negotiation processes in their friendship groups and their social networks. In addition to the group interviews with the girls and boys from the longitudinal study, the interviewer was able to convince four other boys to take part in the study. The intention was to extend the scope of the study in age and gender. The boys, aged 12, 14 and 15, with lower-level and higher-level education, were contacted on a public place in the city centre and took part in the research project on this one occasion. They were interviewed in one group session several days after the first contact in March 2007. These boys also sent their phone's visual contents to the laptop of the interviewer. In general, the girls and boys from the longitudinal research project as well as the boys who were a single time were rather uninhibited in showing and sending their mobile phone content to the interviewer's device. Only a few

---

[1] In Germany we have different types of school. They differ in duration and intensity of education. The ten years education is not so intense and qualifies for apprenticeship. The twelve years education is very intense and qualifies for academic studies. Hence I use the phrase higher-level education to refer to the twelve years education.

[2] Infrared and Bluetooth are wireless and free phone-to-phone transmission functions. The Bluetooth function is especially important to transmit large files like video clips in a short time.

pictures were not sent to the interviewer (e.g., one boy was very reluctant to send a picture of a girl he got to know on the instant messenger program "ICQ").

In summary, the results presented in the next chapter are based on 404 pictures, animations and video clips and attempt to answer the following questions:

- What are the topics reflected in the pictures, animations and video clips?
- What can be said about the origin and distribution of this content?
- How do the girls and boys evaluate and use this content, and for what reasons?
- Are there any interesting differences among the adolescents related to age, sex and education?

The interpretation of the empirical data was carried out through reconstructive methods using Grounded Theory (Glaser & Strauss, 1998; Krotz, 2005b). The aim was to extract categories reflecting the perspective of the adolescents and the content of the pictures, animations and video clips. For this reason, the group interviews were taped and transcribed. In a second step, the interviews and the mobile phone content were structured using Hyperresearch qualitative analysis software. Using the theoretical background of mediatization and social relationships in adolescence, the following categories represent the main results from the analysis of visual mobile phone content.

3.1 Everyday-Life Situations

First, the material revealed "Everyday-Life Situations" as one of the most important topics. The girls and boys from age 12 to 17 used the camera function of their mobile phones to record positive experiences with their friends during shared activities. They also took pictures of their pets, their own room, their bike or their hometown. These pictures and video clips are used in a very intimate way, meaning that the adolescents watched this content when they felt bored or lonely. Furthermore, they shared the content with close peers via the Bluetooth function of their mobile phone, but they do not send it to others. Sometimes they uploaded pictures to personal profiles such as social network sites on the internet.

3.2 Favorite Media Content

The second topic deals with "Favorite Media Content." The adolescents most often saved pictures of music stars on their mobile phones, but they also saved clips from movies or television. The girls and boys downloaded this content from web sites for free or from their peers via Bluetooth. Typical sharing situations are during school, while waiting together at the bus stop or while spending time together at a party. No picture or clip observed originated from a commercial provider. The adolescents gave two reasons for this fact. First, commercial content is too expensive, and they are not willing to pay for it. Second, it is very important for them to exchange the clips and pictures via Bluetooth with friends and peers, and commercial content often prevents such transfers. Thus, commercial content becomes unattractive. The girls and boys used their favorite media content to express their interests and preferences and to talk about it with best friends and peers.

3.3 Sexuality

The third important topic concerns "Sexuality" on different levels. On the first level, the adolescents, especially the girls at the age of sixteen, dealt with their gender identity and sexuality. They took pictures of themselves with different make-up, in underwear or in different outfits, presenting their femininity. But the girls also took pictures of the boys from their social network who they are interested in or took pictures of their current or ex-boyfriends. Sometimes the girls took these pictures covertly, for example, during a party. In some instances, the boys strike erotic poses consciously to please the crowd and appeal to the girls. By this means, the girls tried to document and reflect current developments concerning gender identity and partnership. They kept this content very private, showing it only to their best friends, and they do not send it to anybody else.

On the second level, the girls and boys dealt with sexuality in a funny, provocative and stereotypical way. The pictures and animations show well-known comic figures, like Garfield or Homer Simpson, in explicit activities, or include statements with allusions to sexuality. The main interest of the adolescents in this context was to collect and share the content and to present it to peers to joke, to provoke or to attract attention. The older boys collected pictures revealing stereotypical scenes, like a well-proportioned fair-haired girl in a bikini, lolling on a red Lamborghini. The funny and stereotypical content was not bought from commercial providers, but exchanged via Bluetooth and downloaded from free websites, mostly via the search engine Google.

The third level includes pornographic video clips of typical, explicit scenes similar to a porn film as well as scenes of more or less extreme activities (for instance, between two very corpulent persons) contextualized with funny or dramatic background music. Only the boys – but including all ages and education levels – dealt with these kinds of video clips. They received this content from other, often older, friends from school or leisure activities. The older boys were interested in and curious about this content because it touches new and forbidden topics, and it is common practice to save this kind of content on the mobile phone for a length of time. After some time, they lost interest in such video clips and deleted them. The younger boys said that they shared this content because they wanted to have what the others have. Nevertheless, they have an ambivalent perspective on this. The younger boys argue that the clips are funny to watch with friends, but they also acknowledge that they are sometimes disturbing to watch. They want to collect these video clips but do not like to watch them.

3.4 Violence

The last topic refers to different levels of violent content. Girls and boys of every age and every education level collected and shared funny animated clips where, for instance, comic figures like Tom & Jerry fight each other.

Psychological violence includes pictures and video clips taken in private and awkward situations. Typical examples are covertly-made tapes of teachers or classmates in school. The boys produced and shared this type of video clip. The aim was to laugh about somebody when the clip is exchanged with others. Each adolescent wanted to become the owner of the coolest video clip in the peer group.

The pictures and video clips of physical violence show animals and human beings being beaten, tortured and killed, including car accidents. Again, the boys got this content from their peers via Bluetooth, but they cannot say who the original owner or producer was. The more realistic a picture or video clip seems to be, the more popular and interesting it was for them. The boys watched it together and loved the provocation and the thrill that goes along with it. Like with pornographic content, the younger boys had to force themselves to watch extremely violent content.

At least one additional topic should be mentioned, even though it did not play an important role in this study, as it may become important in the future. Another topic of content exchanged portrayed political statements and ideologies. This content type is also shared with peers and could create new avenues of political participation among adolescents. However, this content also provides problematic political organizations with the opportunity to reach young people and propagate ideologies in a very easy and direct way.

4 Conclusions

The empirical results of this study confirm that the investigation of communicative practices with visual mobile phone content must take into consideration specific relationships and their negotiated meanings. Those negotiation processes are embedded in the everyday life contexts and especially in developmental challenges that are typical for adolescence. The thirteen participants in this study use visual mobile phone content to save everyday life situations and to reflect their own gender development, but mainly to negotiate relationships. The adolescents, especially the girls, share intimate content with their best friends to sustain and deepen these friendships. To reflect their relationship to potential or lost partners, they take and collect pictures from them. The girls and boys try to test and break boundaries and to demarcate their communicative practices from the adult world by sharing and collecting provocative animations and video clips. The boys are also interested in sexual and violent content because they want to be as cool as their friends. After observing the communication content, it becomes clear that the person who is able to acquire the funniest, most embarrassing or even the most violent pictures or video clips is considered the coolest or most admired member of the peer group. This fact is most relevant for the boys of the study.

When we look at the process of communication itself, it appears that the interviewed adolescents are able to obtain almost everything they want from another mobile phone owner. Therefore, getting most out of the pictures and clips subsequently means passing it on to the next person in the group. Thus, communication with mobile phone content can be described as an important practice of participation that is not confined to the communication between two people. The sharing practices are directed toward the integration of one person into a best friends group or into a network of relationships. In doing so, the process of sharing itself seems to be more important than the content that is shared, which means that participating in the analyzed peer relationships means participating in a dynamic and continuous exchange of mobile content. The consequence is that the two girls who own a mobile phone without the Bluetooth features are excluded from this important practice that links peer-relations. During the interview sessions both were very unhappy that they could not join the conversation in this way and felt neglected. Furthermore, they were angry at their parents because they do not support their intention to buy a new multifunctional mobile phone with a Bluetooth feature.

The more practical study conclusions deal with implications surrounding the use of problematic mobile phone content during adolescence. From the investigation of other media, especially television, it is known that the reception of violent content such as horror films, reality shows or news magazines can lead to an emotional overload in childhood and adolescence and can have an a negative effect on the development of one's worldview and the steps one can take to solve conflicts. This media mainly includes clips of physical violence, but also psychological and structural violence (von Feilitzen, 2009; Kunczik & Zipfel, 2004; Murray, 2003; Theunert, 1996). Furthermore, there is a difference between the *reception* of violent media content and one's *own violent activities* that are linked to media content. Adolescents watch violent movies because they are curious to see unknown and forbidden content. They use it for trials of courage, to gain status and prestige, and to test and break boundaries. Only those adolescents that grow up in a problematic social context (e.g., living in a problematic family, having no friends, being outsiders in school) and do not have any other strategy to solve conflicts use violent media content as a valve or as role model to deal with their problems (Grimm & Rhein, 2007; Schell, 2007).

These findings about the use of problematic and violent mobile phone content must be updated and reconsidered because of at least four reasons. The first reason refers to *equipment and access*. Nearly every adolescent owns a mobile phone with audiovisual features and transmission functions, and unlike TV sets or personal computers, they carry their devices with them constantly and have instant access to them. Second, it is not only possible to watch problematic media content with the mobile phone but also to easily *produce and distribute* this content without needing money or any special knowledge or equipment. Furthermore, it is very easy to publish private media content from the mobile phone on other media like TV or the Internet by sending multimedia messages to music programs or by uploading pictures or video clips to personal profiles on social network sites. The third reason alludes to the changed *inhibitions* of using problematic media content. Because virtually everybody owns a portable and multifunctional mobile phone, there are nearly no temporal, spatial, technological, financial or educational barriers to the use of this material. Hence, because of the Bluetooth function, every mobile phone owner, even children, can be considered potential recipients and users of pictures or video clips dealing with pornography, violence or problematic political content. It only takes one click to get what somebody wants and when he or she wants it. Furthermore, it is very easy to take a picture or to film somebody and to upload or download material onto or from the internet, which leads to the fourth aspect of *traceability and control*. It is very difficult or even impossible to determine the identity of the original producer or first sender of problematic or violent mobile phone content. These four reasons emphasize that there are new ways of using violent media content. In particular, more subtle but also very extreme forms of violence are connected to the mobile phone and can have serious consequences for adolescent development. For example, a person may laugh at or take revenge upon somebody using a mobile phone by filming him or her in a private or awkward situations and publishing the content. Thus, the use of mobile phones during adolescence poses new questions about the legal protection of minors and the encouragement of media literacy. The stabilization of a critical use of new communication technologies becomes more and more important not only for children and adolescents but also for parents and teachers as well as providers of technologies and services.

In summary, the mediatization of social networks in adolescence is extended by communicative practices with the mobile phone and especially with visual mobile phone content. In this regard, mediatized communication is realized against the background of developmental challenges and is intrinsically tied to face-to-face-communication and other media. This mediatized intersection of negotiation practices in the relationships of adolescence goes far beyond single, technology-centered media effects. In fact, these processes have to be understood as a comprehensive mediatization of cultural practices involving the change of processes and results of socialization in adolescence.

References

Barthelmes, J. & Sander, E. (1997). Medien in Familie und Peer-Group: Vom Nutzen der Medien für 13- und 14-jährige. Medienerfahrungen von Jugendlichen. München: DJI-Verlag

Blumer, H. (1969). *Symbolic Interactionism: Perspective and Method*. Berkley, Los Angeles: University of California Press.

Castells, M., Fernández-Ardèvol, M., Qiu, J., & Sey, A. (2007). *Mobile Communication and Society: A Global Perspective*. Cambridge, Massachusetts, London: MIT Press.

Certeau, M. de (1988). *The Practice of Everyday Life*. Berkley, Los Angeles: University of California Press. (Original work published 1980)

Corsaro, W. A. & Eder, D. (1990). Children's Peer Cultures. *Annual Review of Sociology, 16*(1), 197-220.

Feilitzen, C. von (2009). *Influences of Mediated Violence: A Brief Research Summary*. Göteborg: Nordicom.

Feldhaus, M. (2004). *Mobile Kommunikation im Familiensystem: Zu den Chancen und Risiken mobiler Kommunikation für das familiale Zusammenleben*. Würzburg: Ergon Verlag.

Fine, G. A. & Kleinman, S. (1983). Network and Meaning: An Interactionist Approach to Structure. *Symbolic Interaction, 6*(1), 97-110.

Furman, W. & Simon, V. A. (1998). Advice from Youth: Some Lessons from the Study of Adolescent Relationships. *Journal of Social and Personal Relationships, 15*(6), 723-739.

Gille, M., Sardei-Biermann, S., Gaiser, W., & de Rijke, J. (2006). *Jugendliche und junge Erwachsene in Deutschland: Lebensverhältnisse, Werte und gesellschaftliche Beteiligung 12- bis 29-Jähriger*. Wiesbaden: VS.

Glaser, B. G., & Strauss, A. L. (1998). *Grounded Theory: Strategien qualitativer Forschung*. Bern: Hans Huber.

Grimm, P. & Rhein, S. (2007). *Slapping, Bullying, Snuffing! Zur Problematik von gewalthaltigen und pornografischen Videoclips auf Mobiltelefonen von Jugendlichen*. Berlin: Vistas.

Haddon, L. (2004). *Information And Communication Technologies In Everday Life: A Consice Introduction And Research Guide*. Oxford: Berg.

Havighurst, R. J. (1972). *Developmental Tasks and Education*. New York: McKay.

Ito, M., Horst, H., Bittanti, M., Boyd, D., Herr-Stephenson, B., Lange, P. G., Pascoe, C.J., & Robinson, L. (2008). Living and Learning with New Media: Summary of Findings from the Digital Youth Project. *The John D. and Catherine T. MacArthur Foundation Reports on Digital Media and Learning*.

Kolip, P. (1993). *Freundschaften im Jugendalter. Der Beitrag sozialer Netzwerke zur Problembewältigung*. Weinheim, München: Juventa.

Kraut, R., Brynin, M., & Kiesler, S. (Eds.) (2006). *Computers, Phones and the Internet. Domesticating Information Technology*. New York: Oxford University Press.

Krotz, F. (2001). *Die Mediatisierung kommunikativen Handelns. Der Wandel von Alltag und sozialen Beziehungen, Kultur und Gesellschaft durch die Medien*. Wiesbaden: Westdeutscher Verlag.

Krotz, F. (2005a). Mobile Communication, the Internet and the Net of Social Relations: A Theoretical Framework. In K. Nyiri (Ed.), *A Sense of Place* (pp. 447-457). Vienna: Passagen Verlag.

Krotz, F. (2005b). *Neue Theorien entwickeln: Eine Einführung in die Grounded Theory, die Heuristische Sozialforschung und die Ethnographie anhand von Beispielen aus der Kommunikationsforschung*. Köln: Halem.

Krotz, F. (2007). The meta-process of 'mediatization' as a conceptual frame. *Global Media and Communication, 3*(3), 256-260.

Krotz, F. (2008). Computerspiele als neuer Kommunikationstypus. Interaktive Kommunikation als Zugang zu komplexen Welten. In T. Quandt, J. Wimmer & J. Wolling (Eds.), *Die Computerspieler: Studien zur Nutzung von Computergames* (pp. 25-40). Wiesbaden: VS.

Krotz, F. & Schulz, I. (2006). Vom mobilen Telefon zum kommunikativen Begleiter in neu interpretierten Realitäten: Die Bedeutung des Mobiltelefons in Alltag, Kultur und Gesellschaft. *Ästhetik & Kommunikation, 37*(135), 59-65.

Kunczik, M. & Zipfel, A. (2004). Medien und Gewalt. Befunde der Forschung seit 1998. Research paper for the BMFSFJ (Bundesministerium für Familie, Senioren, Frauen und Jugend). Available: http://www.bundespruefstelle.de/bmfsfj/generator/bpjm/redaktion/PDF-Anlagen/medien-gewalt-befunde-der-forschung-sachbericht-langfassung,property=pdf,bereich=bpjm,sprache=de,rwb=true.pdf [May 25, 2011].

Lenz, K. (1989). *Jugendliche heute: Lebenslagen, Lebensbewältigung und Lebenspläne.* Linz: Veritas.

Ling, R. (2004). *The Mobile Connection: The Cellphone's Impact On Society.* Amsterdam: Morgan Kaufmann.

Ling, R. (2007). Children, Youth, And Mobile Communication. *Journal of Children and Media, 1*(1), 60-67.

Livingstone, S. (2002). *Young People and New Media: Childhood and the Changing Media Environment.* London: Sage.

Medienpädagogischer Forschungsverbund Südwest (MPFS) (2009). JIM-Studie 2009: Jugend, Information, (Multi-) Media. Basisuntersuchung zum Medienumgang 12- bis 19-Jähriger. Available: http://www.mpfs.de/fileadmin/JIM-pdf09/JIM-Studie 2009.pdf [May 25, 2011].

Murray, J. P. (2003). The violent face of television: 50 years of research and controversy. In E. L. Palmer & B. M. Young (Eds.), *The Faces of Televisual Media: Teaching, Violence, Selling to Children* (pp. 143-160). Mahwa, New Jersey: Lawrence Erlbaum.

Oerter, R. & Dreher, E. (1998). Jugendalter. In R. Oerter & L. Montada (Eds.), *Entwicklungspsychologie: Ein Lehrbuch* (pp. 310-395). Weinheim: Psychologie Verlags Union.

Schell, F. (2007). Jugendmedium Handy: Motive und Problemlagen im Zusammenhang mit der Nutzung gewalthaltiger und pornografischer Inhalte. Available: http://www.msa.jff.de/dateien/Motive_und_Problemlagen.pdf. [February 27, 2007]

Schuh, R. F. (2007). *Die mobile Generation: Jugendliche und ihr Handy.* Saarbrücken: VDM Verlag.

Shell Deutschland Holding (Ed.) (2006). *Jugend 2006: Eine pragmatische Generation unter Druck.* Frankfurt am Main: Fischer Taschenbuch.

Suoninen, A. (2001). The Role of Media in Peer-Group Relation. In S. Livingstone & M. Bovill (Eds.), *Children and their Changing Media Environment: A Comparative European Study* (pp. 201-219). Mawah, New York: Erlbaum.

Tapscott, D. (2009). *Grown up digital: How the net generation is changing your world.* New York: McGraw Hill.

Taylor, A. S. & Harper, R. (2003). The gift of the gab? A design oriented sociology of young people's use of 'mobilZe!'. *Journal of Computer Supported Cooperative Work, 12*(3), 267-296.

Theunert, H. (1996). *Gewalt in den Medien – Gewalt in der Realität: Gesellschaftliche Zusammenhänge und pädagogisches Handeln.* Opladen: Leske + Budrich.

Zinnecker, J., Behnken, I., Maschke, S., & Stecher, L. (2002). *Null Zoff & Voll Busy: Die erste Jugendeneration des neuen Jahrhunderts.* Opladen: Leske + Budrich.

Celebration and Concern

Digitization, Camera Phones and the Citizen-Photographer

Frank Möller

1   Introduction: Photography in a Pixellated Age

We live in what Danchev (2009, p. 71) calls a "pixellated age". Digitization has changed how people communicate with each other, as is nicely illustrated in the following episode from a Michael Dibdin book (1999, p. 288):

> "Give me your mobile," the other man told Zen.
> "I don't have one."
> The man stared at Zen in total disbelief.
> "Well, actually I do", Zen went on, realizing that he was cutting a poor figure.
> "But I left it at home. I never use it, to be honest. The last thing I want is people being able to get in touch with me day or night, wherever I may be. I'm [sic] suppose I'm old-fashioned."
> Nello laughed.
> "You're not just old-fashioned, Papà. You're extinct!"

Digitization has also changed, perhaps even revolutionized, the techniques of producing, altering and disseminating images. According to Ritchin (2009, p.11), some 250 billion digital photographs were taken in 2007, and "nearly a billion camera phones were said to be in use". Ritchin (2009, p. 126) also notes that the worldwide reach of the Internet "is a source of both celebration and concern". Likewise, the worldwide production and dissemination of digital imagery is similarly a source of celebration and concern. Is the pixellated world a better place? Or is it just a different place?

The digital age is different from the analog age because, among other reasons, many people who formerly would not have thought of taking pictures now take them as a matter of course. To adapt the episode that opened this chapter to the age of the camera phone: The individual who does not own or use a camera phone is not merely old-fashioned but (almost) extinct, at least in the image-obsessed Western world of multimedia. Questions pertaining to the ethical responsibility of the photographer, which were previously relevant only to a relatively small group of people, affect many more people when everyone is a photographer. No one equipped with a digital camera can avoid such questions when taking photographs of other people, thereby treating them as subjects of one's photography. Further, no one looking at images of other people can avoid questions pertaining to the responsibility of the spectator.

The pixellated age is different also because digitally produced images and especially pictures taken by camera phones can be disseminated much faster and wider than analog photographs by submitting them to diverse online communities. From these virtual networks, photographs often develop their own dynamics, which are uncontrolled by the photographer, for example, by becoming a part of another person's blog as an image that was either copied or altered. This acceleration of images, that is to say, the increase in the number of images combined with worldwide access to them in near real-time, potentially transforms common spectatorship into global spectatorship. It helps develop civil society into a global society, thus transforming what Azoulay (2008) calls the citizenry of photography into a truly global, de-territorialized citizenry that extends beyond the boundaries of nation-states. However, the implications of these developments have not been sufficiently examined. We tend to think, talk and write about digital photography in terms and modes suggested and informed by the literature on analog photography – critical as well as affirmative, skeptical as well as idealizing, and thus in terms other than its own. There is a marked paucity of literature on such contemporary and important trends as the introduction and widespread use of camera phones while every other day new books are published and exhibitions organized on the classics of photography (Greenough, 2009; Sire, 2009). Books such as Grey's (2009) and Ritchin's (2009) and exhibitions such as *Exposed: Voyeurism, Surveillance and the Camera*[1] are exceptions.

Digital photography is often said to have a different and more problematic relationship to reality than analog photography. However, the relationship of photography to any prior reality has been and always will be problematic. As Simmonds (2009) has argued, photographs are always pre-edited by the photographer prior to the act of actually taking a photograph; they are edited during the process of taking a photograph by selecting the frame (the question of inclusion versus exclusion), the lens, the depth of field and so on. Photographs are also edited by the photographer or a photo editor in the process of preparing a given photograph for publication, including publication on the Internet. Thus, regardless of whether they are analog or digital, photographs are always fabrications.

---

[1]   Tate Modern, London, May 28–October 3, 2010; San Francisco Museum of Modern Art, October 30, 2010–April 17, 2011; Walker Art Center, Minneapolis, May 21–September 11, 2011 (see Phillips, 2010).

Of course, there are substantial differences between analog and digital photography. Ritchin (2009, p. 141) concludes that digital photography "represents an essentially different approach than does analog photography". However, the degree of authenticity has arguably been over-emphasized as a difference between analog and digital. This is not to say that digital photography is more authentic than it is normally alleged to be but instead that analog photography is less credible than it is usually said to be. Thus, it can be argued that our concern over the authenticity and credibility of digital photography including camera phone images is due to an overestimation of the authenticity and credibility of analog photography. We overestimate the authenticity of analog photography not only owing to the powerful tradition of photojournalism and documentary photography (which has a long tradition of claiming that what we see is "real"), but also because in a rapidly changing world, we seek some degree of assurance. However, as identified by numerous writers, such assurance is a myth. If we were aware of this myth, the "public laments about the […] loss of authority and truth" associated with "the increased capacity for pictorial manipulation arising from the use of digital cameras and computer imaging" (Campbell, 2003, p. 65) would perhaps be less frequent.

As Danchev (2009, p. 36) suggests, analog photographs are "not merely illustrations of what was already known. They are new knowledge". They show us something we would not know without them. Otherwise, we would not need them. Digital photographs are also new knowledge. They too show us something we would not know without them (even if this "something" is different from the "something" communicated by analog photography). This knowledge cannot simply be dismissed with reference to digital photography's lack of an original, which renders difficult the verification or falsification of a digital image. Even without verification, digital photographs are new knowledge.

2   "We're Photographers, Not Terrorists"[2]

Owing to the ubiquitous presence and use of camera phones to take pictures and disseminate them in near real-time, the visual construction of political space and the democratization of image-making seem to have reached a new stage. This stage was anticipated by the introduction of digital cameras, which offer new possibilities for the construction of visual space including visual opposition that are beyond governmental control. The triangle that connects the photographer,

---

[2]   "We're photographers, not terrorists," *Guardian*, (12/15/2009). Available online: http://www.guardian.co.uk/commentisfree/libertycentral/2009/dec/11/photographers-section-44 -terrorism-act [December 15, 2009].

the subject and the spectator has become tighter. In visually constructed political space (which will be discussed in greater detail below), there is a compulsion to depict, a compulsion to be depicted and a compulsion to look at the depictions of others if one does not want to exclude oneself, or to be excluded by others, from the realm of the political, incapacitated from acting politically.

In some cases, the non-professional citizen-photographer appears to have replaced the professional photographer as the most important producer of images because professional photographers cannot be in all places at all times. The citizen-photographer can be understood as someone who takes pictures not only for private consumption but to apply his or her politically educated view to photography to inform and educate others politically, thus helping to construct a visual/virtual political community. Even if professional photographers happen to be at the right place at the right time, they might choose different subjects than a non-professional photographer, or they might choose the same subjects but photograph them differently. In addition, professional photographers find it increasingly difficult to go about their business. Based on the re-designation of public space as what has been called "state space"[3] in the journalistic discourse, governments in such different places as Iran and the United Kingdom seem to be interested in suppressing the production of images. Indeed, photographers and the visual construction of political space are not always welcome. An increase in the use of CCTV surveillance techniques can be observed in ostensibly liberal societies, primarily in metropolitan areas. Photography bans are often related to post-9/11 policies focusing on security and are accompanied by omnipresent police officers and private security personnel with obscure accountability. As has been noted in the journalistic discourse on surveillance, these trends endanger every person who "engage[s] in the act of photography in a public place",[4] albeit to different degrees.

Governments might be able to inhibit the work of professional journalists. Restricting the work of citizen-photographers using camera phones for the production and dissemination of images is a different thing entirely, though. Camera phones are vehicles for both instant photography and the instant dissemination of images. As shown by the production and dissemination of the notorious photographs taken at Abu Ghraib prison, digital photography is indeed difficult to control. In particular, photographs and videos taken and disseminated by means of camera phones are capable of upsetting those in power, as illustrated by the video of the death of the student Neda Agha-Soltan during the turmoil in Iran in

---

[3]   "Protecting the media from the police," *Guardian*, (07/01/2010). Available online: http://www.guardian.co.uk/commentisfree/henryporter/2010/jan/07/police-photography-public-space [January 8, 2010].

[4]   "We're photographers, not terrorists," see note 2.

2009. The international response to these images transformed the student into a symbol and icon of the Iranian democratic movement.[5]

Authorities' lack of controllability and the fear of the international response to images (and the conditions depicted therein) help explain the lack of proportionality often underlying their disproportionate response to digital photography. This disproportionality was reportedly acknowledged by Scotland Yard's assistant commissioner for special operations John Yates, who said that there is "'an enormous amount of concern' about the use of anti-terror laws against people taking photographs in the street."[6] From the point of view of the authorities, photographs may be dangerous. From the point of view of citizens, they may be liberating. From the point of view of the subjects of photography, they may be both liberating and oppressive. The critical function of images of people exposed to maltreatment by authorities, for example, is hard to separate from the oppressive function of the same images once they have been made available on the Internet, where they contribute to the visual exploitation of those depicted. For example, whatever subject positions Neda Agha-Soltan inhabited during her short life, she is now reduced to a dying student and thus necessarily misrepresented.

3 Photography and the Approximate

The introduction of digital photography and especially camera phones suggests differentiating between ad hoc, spontaneous photography and chronic photography.[7] The former is the domain of the amateur photographer (and selected photojournalists who happen to be at the right place at the right time), and the latter is the domain of the professional documentary photographer (for example, the photojournalist who is embedded with armed forces in war, the documentary photographer on a long-term assignment or the police photographer who monitors citizens on a regular basis) and the dedicated amateur. Another way to frame

---

[5]   See "How Neda Soltani became the face of Iran's struggle," *Guardian*, (22/06/2009). Available online: http://www.guardian.co.uk/world/2009/jun/22/neda-soltani-death-iran [June 24, 2009]. For the mix-up of names and people, see "Mistaken as an Iranian Martyr, Then Hounded," *The New York Times*, (31/07/2010). Available online: http://www.nytimes.com/ 2010/08/01/world/middleeast/01neda.html?_r= 1&th&emc=th [October 7, 2010].

[6]   "Scotland Yard warns police officers over photography concerns," *Guardian*, (15/12/2009). Available online: http://www.guardian.co.uk/uk/2009/dec/15/yates-police-terrorism-powers-photography [December 15, 2009].

[7]   See also Azoulay's distinction between disasters and chronic disasters (2008, p. 51).

this difference is to distinguish between "the instant" and "the reflective",[8] as has been done in the journalistic discourse on photography (although the instant can also be a result of reflection and reflection can result in instant images). Camera phones seem primarily to be vehicles for ad hoc, spontaneous, instant photography.

Whether ad hoc or chronic, photographs are approximations. As representations, they are never identical with that which they represent. Rather than simply reflecting reality, they construct reality, or better: realities. They carry with them what King (2003, p. 180) has called a "surplus of meaning", that is to say, a plurality of meanings that can be suppressed by means of language but not erased altogether. Is the irreducibility of photographs a merit or a liability? If it is primarily seen as a liability, then the difficulty of authenticating digital imagery by verbal or other means would seem to be especially problematic. However, Foucault (1994, p. 9) has argued that "it is in vain that we say what we see; what we see never resides in what we say" and MacDougall (1998, p. 246) has shown that "[p]ictures and writing produce two quite different accounts of human existence". Thus, writing or talking about images is always problematic because we do not have appropriate language to transform the seeable into the sayable. Trying to do so means approaching images in terms other than their own and limiting them and the knowledge they produce to that which can be said in (a given) language.

Indeed, uneasiness about the approximate, that is to say, uneasiness that we necessarily fail to say accurately what we see, is widespread. Many people seem to find it difficult or politically undesirable to accept that there is something inherently elusive in images that we cannot grasp by means of language no matter how hard we try. For example, Benjamin (1963b, p. 64), writing in the interwar period, suggested elevating the inscription to the most essential ingredient of a photograph; otherwise, photographic construction is bound to "get stuck in the approximate".[9] According to Benjamin (1963a, p. 21), captions give directives to those looking at pictures. Perhaps surprisingly, whether these directives are right or wrong is not important. Regardless of their content, such directives seem to enhance both the readability of photography (they tell the readers what they are supposed to see) and its applicability to political struggle. The distinction between "right" and "wrong" directives is difficult to verify and ultimately irrelevant as long as the directives serve their political purpose. The work of art in the

---

[8]   See "The mafia and me: Mimi Mollica's portraits of Sicilian society," *Guardian*, (14/12/2009).
        Available online: http://www.guardian.co.uk/artanddesign/2009/dec/14/mafia-mimi-mollica-
        photographs [December 14, 2009].
[9]   "[Ohne Beschriftung] (muß) alle photographische Konstruktion im Ungefähren stecken
        bleiben."

age of technological reproducibility thus becomes the work of art in the age of political reducibility. This is perfectly understandable of course in the light of Benjamin's (1963a, p. 44) suggestion to politicize the arts to counter the aestheticization of politics in Fascism (see also Reichel, 1991). Here, the approximate is seen as a liability and not as an asset. The interpretive strategy is to reduce the approximate and not to accept it and build upon it.

This understanding is also at the core of photojournalism and documentary photography (the occasional photo essay being the exception confirming the rule): text seems to explain what we see; the image seems to support the text; and together, text and image create what Gilgen (2003, p. 55) has called a mutually supportive "intellectual stereoscopic effect" that strengthens the overall message or makes the message possible in the first place. This effect seems to work even though many people are aware of the mismatch between verbal and visual constructions of reality. There is no reason to assume a priori that it would not work in connection with digital constructions of reality. The lack of an original renders verification difficult, which seems to make digital photography less authentic and easier to manipulate than analog photography. However, the average viewer does not normally compare a published analog photograph with the contact prints to verify and contextualize the published print. Further, given the volume of digital photographs nowadays normally taken of any major or even minor event, digital photographers and digital photographs mutually control one another to some extent. The technically skilled computer user can be both a manipulator and a watchdog.

Thus, there will always be a residue of uncertainty in connection with digital photography, but this may not be entirely negative. If we understand the approximate as an asset that can teach us to live with, appreciate and capitalize on difference, then the potentialities of the irreducibility of digital photography would seem to be obvious. As Couldry (2000, p. 21-22) has noted, working with difference instead of reducing difference is at the core of the intellectual and political commitment of globalized cultural studies.[10] Words, however, limit images to that which can be said in a given language in a given situation. Such limitations cannot be politically neutral. Indeed, Benjamin's "directives" are essentially political. The social processes through which an image's various meanings are reduced to one binding and non-negotiable meaning, to *the* meaning of an image, are eminently political processes that cannot be separated from the power relations operating within a given society. Such processes normally confirm rather than challenge the established discursive patterns and power rela-

---

[10]   Of course, not all forms of difference are appreciated here. For example, the elimination of differences in economic performance, health levels, life expectancy or gender roles may very well be appreciated by those who otherwise appreciate difference.

tions that generate, regulate and benefit from these very patterns. Ultimately, they destroy images as a potential source of alternative knowledge production; images become invisible. If we are interested in alternative knowledge production, in the diversification of worldviews, in new perspectives and in what Shapiro (2009, p. 33) calls visual "counter-spaces" that are beyond and unaffected by governmental control, then the approximate would appear to be an asset rather than a liability. It would offer space to resist established representational strategies and would challenge these strategies by presenting alternatives. If we manage to live with the approximate rather than to reduce it, then our obsession with the credibility and authenticity of photography and our uneasiness about digital photography may be replaced by an approach that appreciates the potentialities of digitization to, among other things, help transform common spectatorship into global spectatorship.

4   From Common Spectatorship to a Global Visual Community

In their work on photojournalistic images, Hariman and Lucaites (2007) analyze among other things the social processes through which selected photographs came to be regarded as icons in the national public culture of the United States of America. They are interested in the question of how these images influence the construction of political space. They conclude that the "daily stream of photojournalistic images [...] defines the public through an act of common spectatorship" (ibid., p. 42). The individual viewer can respond politically to an image not as an individual but only as a member of the discursively organized public participating in collective action in response to the image. We live in what Mitchell (1994, p. 2) called a culture that is "dominated by pictures, visual simulations, stereotypes, illusions, copies, reproductions, imitations, and fantasies". In such a culture, the public would not be a public, and the individual, who is powerful only as a part of discursive-collective action, could not exert much political influence without simultaneously viewing images individually and collectively, as a member of the public. The act of viewing constitutes the public, and it is only as a part of the public that the individual can exert political power. By ignoring images, people would position themselves outside the realm of the political and be deprived of the possibility to act politically.

    Hariman and Lucaites apply a Habermasian discursive action approach to images combined with an Arendtian understanding of power where power can only be exerted in cooperation with other people. As such, their approach is an important contribution to both the literature on the discursive construction of political space and the literature on political participation. Three things may be added. First, there is no reason to limit the discussion to photojournalistic icons such as Dorothea Lange's "Migrant Mother" or Nick Ut's "Accidental Napalm". Such icons are increasingly supplemented with and, perhaps, replaced by contemporary digitally produced images, both ad hoc and chronic, including images produced by means of camera phones. For example, Borenstein (2009) has shown how the camera phone images taken during the 2005 London bombings shaped the ways news is produced (see Figure 1 below), and the BBC regularly invites viewers to submit their own images, many of which are taken by means of camera phones. The limitation to photojournalistic icons is indicative of the historical, almost nostalgic, approach to photography in certain strands of the literature that often laments what is lost in the process of digitization and fails to grasp what is gained. Second, there is no need to limit the discussion to common spectatorship: the visual construction of political space requires producers, subjects and spectators of images. The relationship between these different subject positions is of utmost importance because it is through this relationship that the various possible meanings of a given image are negotiated. Third, there is no reason to limit the debate to the national culture of the United States: common spectatorship increasingly means global spectatorship, and common visual space increasingly means global visual space. The images produced in lower Manhattan on September 11, 2001 exemplify this in that they transformed a local event into a global event.[11] The speed with which images travel from one place to another has increased tremendously in the digital age; the frequent use of camera phones for the production and dissemination of images has contributed a great deal to the acceleration and the limitlessness of images.

Indeed, within certain limits, the digital age is a global age where individuals can connect with other producers, spectators and subjects of photography on a global scale. The question of who gets heard has been an essential component of cultural research for quite some time. The obvious follow-up question would be the question of who gets seen. The absence of visual representation contributes to the marginalization and exclusion from the political of groups of people who literally become invisible. Every process of marginalization reflects unequal power relations. Visual representations of other people also reflect power relations between the photographer and his or her subject, and the prioritization of visual culture marginalizes those groups of people who do not share the Western interest in images and subscribe to different sensory hierarchies. Images tend to exploit human beings visually, to freeze subjects (for example, as victims; see above) and to deny them agency. This is frequently acknowledged in visual ethnography, visual anthropology, cultural studies, peace research and

---

[11]   See, for example, the painted narrative scrolls of the Naya women in India thematizing "Bin Laden" and "September 11" (Fruzzetti and Östör, 2007, pp. 88–94).

elsewhere. Couldry (2000, p. 58) has noted that we live "in societies and cultures where individuals are spoken *for*, much more than they speak in their own name", but we also live in societies and cultures where individuals are frequently visually represented by others rather than representing themselves. In both cases, they are not necessarily represented "accurately" (ibid., p. 58), that is, in accordance with their own self-image and the way they would want to be represented.

If we agree that the act of viewing constitutes the public, through which the individual has potential political power, then it would seem to be mandatory to produce images of other people including images of marginalized people and even images of people in pain. Otherwise, human suffering would be positioned outside the realm of the political and it would not be possible to respond to it politically. Thus, regardless of the feelings of the victims, human suffering must be captured visually so that it is not depoliticized. The construction of political space through images implies both the compulsion to look and the compulsion to depict. However, the acts of looking and depicting are likely to conflict with the feelings and interests of the victims when the victims do not represent themselves.

Digital photography and especially camera phones offer many relatively inexpensive possibilities for marginalized groups to exert agency by becoming photographers, thus representing themselves and disseminating their images on a wider scale than ever before. Thus, digital photography may allow marginalized groups to correct the image that viewers have of them. However, this also depends on the viewers' willingness to correct their image of other people. Correcting one's image of other people implies correcting one's image of oneself because the self and the other are inseparably connected and interdependent. Still, photographic self-representations of marginalized people can contribute to the diversification of perspectives on their and our worlds. For example, Ritchin (2009) has observed a marked discrepancy between Western perceptions of postgenocide Rwanda and the self-representations of children at the Imbabazi orphanage in Gisenyi made using first disposable and later digital cameras. These photographs "are much more lively [than official reports], responding to color and light and their neighbors with considerable wonder. They refuse to be the symbol of their people's tragic history" (ibid., p. 127). By representing themselves, marginalized people contribute to the discursive construction of global political space and they do so not as objects of another's photography but as agents of their own image.

5 Photojournalism in the Pixellated Age

Representations of wars, famines, accidents, terrorist attacks and other forms of human suffering that aestheticize their subject are often accused of depoliticization. They are alleged to divert the viewer's attention from the conditions depicted in the image to the beauty and design of the image. However, as Strauss (2003) and Bal (2007) have noted, representation necessarily aestheticizes; it cannot avoid doing so. It stylizes; it transforms. It is a representation, after all. The underlying supposition is that less aestheticized, less elaborated photography would help avoid the viewers' depoliticization and increase their degree of empathy.

The one thing most instant, ad hoc photographs taken by means of camera phones are not is elaborate. Their basis is spontaneity and their purpose is the communication of spontaneity. Frequently, their lack of elaboration and reflection is more than compensated for by the degree of authenticity and urgency they manage to communicate to the viewer: nothing is staged and nothing is fabricated; everything is real and authentic – or so it seems; indeed, it seems that what is represented is what really happened. For example, the notorious Abu Ghraib photographs were, as Simpson (2006, p. 104) noted, "accepted as the real thing" because they were immediately recognizable "as the products of amateurs" (ibid., p. 106). Indeed, even the fiercest apologists of what happened at Abu Ghraib have not called into question the authenticity of these images.[12] Robert Capa's famous photographs of the D-day landings in 1944 are powerful precisely because of their technical imperfection (a result of printing errors at the lab in London), which helped communicate "sea-drenched authenticity and unprecedented immediacy".[13] Viewers may be attracted precisely by the seeming lack of compositional elaboration (which, in art photography, may be carefully elaborated indeed).

Ad hoc photographs are also often blurred, unfocused and grainy – one might call them refreshingly unprofessional – but the lack of compositional elaboration is not carefully constructed. They appeal to the viewer despite or even because of this lack. Ad hoc photographs communicate with the viewer. They show the viewer things he or she would not know otherwise; they are new knowledge. They give the viewer the feeling of being at the place and the exact point in time where the photographs were taken. Thus, they intimately involve

---

[12] To be sure, the snappers at Abu Ghraib are not citizen photographers as defined above, that is to say, people who apply their politically educated views to photography to inform and educate others politically.

[13] See "'I was there,'" *Guardian*, (18/10/2008). Available online: http://www.guardian.co.uk/ artanddesign/2008/oct/18/war-photography [October 18, 2008].

the viewer, providing him or her with little room to escape. The lack of an original does not infringe upon their power. They are regarded as documents or as evidence. They are authentic even though they cannot be authenticated. They are credible even though they cannot be verified. They are not art, but they are "artful" in that they are "skillfully," if often intuitively, "adapted for a purpose" (Little, 1973, p. 109). The purpose is to show, to document, and to prove what happened.

The camera phone photographs taken by Alexander Chadwick at the occasion of the 2005 London bombings, for example, are among the small group of digitally produced camera phone images that can be considered iconic, at least in the national cultural context of the United Kingdom. As the Guardian stated in its gallery celebrating 100 years of press photography, they belong to the "key images to have remained in people's consciousness since the 7/7 bombings on London Underground." These "image[s] – taken by someone involved in the disaster – ha[ve] forced a reconsideration of the way in which press images are both produced and circulated in the 21st century".[14] Of course, they have forced us to reconsider many other things as well.

This is ad hoc photojournalism in the pixellated age. It is a new form of photojournalism where the images are captured by someone involved in the incident who is not a professional photographer but whose photographs nevertheless make it to the front pages. Although it is new, this new form is dedicated to photojournalism's traditional principles: to paraphrase Capa's famous dictum (that got him killed in Korea), the photographs are good because the photographer is close enough.[15]

---

[14]   "100 years of great press photographs," Guardian, (10/10/2009). Available online: http://www.guardian.co.uk/artanddesign/gallery/2009/nov/10/100-years-press-photography? picture=355415227 [November 10, 2009].

[15]   Capa (as quoted in Ritchin, 2008, p. 150) said that "If your pictures aren't good enough, you aren't close enough." However, as the recent debate on Capa's Spanish Civil War photography shows, even if you are not close enough, your pictures can be very good indeed.

Figure 1: Alexander Chadwick's camera phone photograph of the 2005 bombings on London Underground; copyright Sipa Press, reproduced with permission.

Closeness is often poignant. It is unnerving to even think about camera phone photographs taken inside the collapsing World Trade Center on September 11, 2001 and disseminated by people who would not survive the collapse of the towers. Likewise, it is unsettling to imagine camera phone images taken by people in an Afghan or Palestinian village and disseminated minutes or even seconds before being killed by a missile, which is also directed by some form of photographic device. Such images would be a photographic archive of the soon-to-be-killed and the soon-to-die, thus creating a virtual/visual community and adding to existing photographic archives of the disappeared in places such as Algeria and Argentine, as analyzed by Downey (2009). Is it possible for photography to be more authentic, more poignant, more urgent, or more "true" than it is in such images?

Such scenarios tremendously complicate the subject positions of viewers who are exposed to images of people in pain. As I have argued elsewhere (Möller, 2009), when exposed to such images, viewers normally face a dilemma wherein neither looking at the images nor ignoring them seems to be an option. On the one hand, ignoring such images would position the individual outside the visually constituted political public, but it is only as a part of this public that he or she can exert political power. On the other hand, looking at such images is believed to prolong the subject's victimization and exploitation. Because representation always comes after the fact, the viewer has little ability to undo that which is depicted in a given image; as Sliwinski (2004, p. 154) has concluded, being "witness to suffering brings with it the demand for a *response*, and yet one's response to photographs can do nothing to alleviate the suffering depicted". This is especially true with respect to images of dead bodies: whatever the response to such images, the dead cannot be revived.

Commenting on Alexander Gardner's photograph of Lewis Paine (also known as Lewis Powell) taken prior to Paine's execution for the attempted assassination of US Secretary of State William H. Seward in 1865, Barthes (2000, p. 96) commented that the punctum of this photograph is that "he is going to die. I read at the same time: This will be and this has been; I observe with horror an anterior future of which death is the stake". In this case, the subject of the photograph, Lewis Paine, was still alive when the picture was taken, but he had been dead for quite some time when Barthes regarded the photo and reflected upon it. In the scenarios outlined above, however, the soon-to-die are not yet dead when viewers see their pictures that are disseminated in real time by camera phones, which greatly complicates the viewers' position, perhaps unbearably so. In this case, it appears to the viewer that there is a window of opportunity (that was absent in Barthes's case) between the moment when the camera phone images were taken, disseminated and looked at and the imminent death of the people depicted: the soon-to-die are not yet dead; they are still alive when we look at their pictures. Because they are still alive, as viewers, we may get the impression (which may very well be an illusion) that there is something we can do. It imparts viewers with the responsibility to prevent their deaths even though there is nothing to be done. The subjects will die, and with them, a part of us will die, too.

6   Irony in the Digital Age

The distinction made by Danchev (2009, pp. 70-73) between seeming and posing is crucial in connection with political communication, especially concerning self-representations of those in power. Such self-representations often carry with them an air of vanity, self-importance and self-aggrandizement that can be and often has been uncovered by caricatures that use irony and exaggeration. The painter and sculptor Fernando Botero, for example, has masterfully depicted the politicians of his native Colombia by, as Ebony (2006, p. 10) notes, "pok[ing] fun at the trumped-up grandeur of the military leadership then in control" while celebrating ordinary women and men. Any attempts to challenge, undermine and reduce to absurdity such self-representations will benefit from taking into consideration the distinction between seeming and posing, which is basically a distinction between the image that a person wants to communicate to others and the image that he or she actually communicates. As Danchev (2009, p. 71) succinctly states, posing is "bad acting. Posing is seeming gone wrong".

Although it may seem old-fashioned, the photographic portrait of political leaders continues to matter to them in the digital age as much as ever. Success and satisfaction are not guaranteed, however. Danchev (2009, p. 70) points out that politics "is a performance" but "[a]uthenticity is an enigma" (ibid., p. 72). To authenticate one's image is necessary but difficult. For example, according to the photographer Nadav Kander, former British Prime Minister Gordon Brown often appeared nervous in pictures not because he was insecure about his policies, but "because he's insecure about his appearance".[16] Insecurity about appearance was communicated and perceived as insecurity about policies, thus undermining political leadership and trust in it. In contrast, people like former US president Ronald Reagan do not normally appear nervous in pictures not because they are confident about their policies but because they are confident about their appearance. This is good acting: security about appearance is communicated and perceived as security about policies. It strengthens political leadership and trust in it; a strong performance is perceived as strong leadership. Other politicians, such as the former British Prime Minister Tony Blair, are confident of both their policies and their appearance but may ultimately fail to convince the audience, perhaps because they are too convinced of their policies, perhaps because self-confidence becomes self-indulgence, or perhaps because self-confidence prevents politicians like Blair from acknowledging the consequences of his or her policies.

---

[16]   Nadav Kander, as quoted in "Follow the leaders: the art of the political portrait," *Guardian*, (03/01/2010). Available online: http://www.guardian.co.uk/artanddesign/2010/jan/03/political-portrait-obama-blair-brown [January 3, 2010].

Figure 2: kennardphillipps: "Photo Op, 2005"; reproduced with permission by
            Peter Kennard and Cat Phillipps.

Peter Kennard and Cat Phillipps's montage of a simpering Tony Blair taking a
picture of himself with what seems to be a camera phone in front of a huge explosion reminiscent of the war-afflicted devastation in Iraq, Pakistan and Af-
ghanistan, is revealing of Blair's autism of power, that is to say, his insuscepti-
bility to anything outside his own belief system.

    Kennard and Phillipps' photograph of people on Oxford Street taking, by
means of camera phones, photographs of the above montage of Blair is even
more striking and more indicative of the potentialities of instant communication
in the pixellated age. These photographs can be disseminated almost instanta-
neously throughout the nation and even the world. Their production and dissemination cannot be controlled or regulated; they cannot be stopped. As such, they
may help stop Tony Blair's policies by communicating to the viewers that which
Blair himself failed to see.

Figure 3: kennardphillipps: "Photo-Op in window Santa's Ghetto, Oxford St, London, 2006"; reproduced with permission by Peter Kennard and Cat Phillipps.

7 Surveillance, Coveillance, Sousveillance

The history of photography can be told as a history of surveillance (cf. Phillips, 2010). As Sontag (1977, p. 5) has noted, photographs were used by "the Paris police in the murderous roundup of Communards" as early as 1871, and they later "became a useful tool of modern states in the surveillance and control of their increasingly mobile populations". If we understand power as an institution that must be controlled, checked and balanced, then the digital age is a source for concern and celebration. On the one hand, it is a source for concern because contemporary societies have become surveillance societies based on digitized administration. Such societies are characterized by "purposeful, routine, systematic and focused attention paid to personal details, for the sake of control, entitlement, management, influence or protection" (Ball & Murakami Wood, 2006, p. 4, § 3.1). As has been observed in journalistic discourse on the surveillance society, especially after September 11, 2001, public space has been increasingly redefined as space "over which the police and CCTV systems have exclusive photographic rights."[17] On the other hand, it is a source for celebration because potential counter-surveillance technologies such as camera phones give citizen-photographers ample opportunities to exert visual opposition to the all-seeing

---

[17]   "Protecting the media from the police", see note 3.

eye of the authorities and to undermine and challenge forms of social control.[18] Exaggerated surveillance generates and, arguably, necessitates sousveillance – a term derived from the French words "sous" (below) and "veiller" (to watch). The term was introduced by Mann, Nolan and Wellman (2003) to describe devices "offering panoptic technologies to help [individuals] to observe those in authority" (p. 332). Sousveillance is a source of concern for the authorities because it cannot easily be brought under control. "Surveillance cameras threaten autonomy" (Mann et al., 2003, p. 347). Sousveillance cameras threaten authority.

However, sousveillance is also a source for both celebration and concern. It can be celebrated as a vehicle people can use to exert visual opposition by taking pictures in situations where it is not formally permitted to take pictures including pictures of those in authority. Such pictures can be taken fortuitously or on a planned and regular basis. Mann et al. (2003) argue that sousveillance is a form of participatory, bottom-up "inquiry-in-performance" (p. 333) that aims to confront bureaucratic organizations and their representatives with the very means these organizations use to observe and control the citizens. At the very least, the public display of sousveillance devices and their performative use is a symbol of non-acquiescence to the omnipresence of surveillance techniques in contemporary surveillance societies. Simultaneously, as Koskela (2004) has suggested, the limited visibility of a camera as a part of a mobile phone and the rotating lens that some models offer[19] limit the visibility of the act of taking a photograph and thus facilitate image production and protect the photographer in situations where photography is not permitted (p. 203).

However, sousveillance is also a source of concern. Taking pictures in situations where it is not permitted to do so can be seen as an act of civil disobedience. Such an act, however, does not automatically or necessarily alter the power relations underlying the taking of pictures and the visual construction of the political. Indeed, authorities backed by powerful law enforcement agencies typically spy on people using CCTV cameras, which are anonymous technological apparatuses that hide the people who operate them and assess the collected data. As Mann et al. (2003) note, sousveillance confronts "individuals using tools to observe the organizational observer" (p. 333) with organizations represented by individual observers and thus describes a profoundly asymmetrical

---

[18]   The topical exhibition *Exposed: Voyeurism, surveillance and the camera* (see note 1) does not differentiate between surveillance and counter-surveillance (see Phillips, 2010, pp. 141–166). Surveillance photographs are said to have in common "a spirit of distance, abstraction, and a certain placid ambiguity" (p. 143). This strikingly unpolitical definition totally abstracts from the politics of both surveillance and opposition.

[19]   The rotating lens is reminiscent of the technique applied by Paul Strand in 1916 when he "fitted his camera with a false lens (the real one was pointed to the side; later he would use a prism)" (Phillips, 2010, p. 20).

relationship. Acts of sousveillance give the impression of equality and symmetry in situations characterized by profoundly unequal and asymmetrical power relations.

Furthermore, although authorities are unable to control sousveillance, they can use it for their own purposes by involving citizens in practices of observation and social control, for example, with respect to border control (cf. Ball & Murakami Wood, 2006, p. 37, § 10.5.5). Indeed, in addition to spying on authorities, sousveillance techniques can also be used to spy on other individuals. In part reflecting a combination of what Andrejevic (2005, p. 482) calls neoliberal "strategies for offloading duties of monitoring onto the populace" and post-9/11 thinking in terms of omnipresent threats and risks, citizens spy on fellow citizens, neighbors on neighbors, workers on co-workers, husbands on wives, wives on husbands, majorities on minorities and so on. Such practices are termed by Mann et al. (2003, p. 338) "*co*veillance". However, this term ignores Eisenman's (2007, p. 99) observation that, normally, "the one who watches is stronger than the one who is watched" even if the one who is watched allows his or her picture to be taken. Thus, the same techniques that can be used for sousveillance in one situation can also be used in another situation for what Andrejevic (2005, p. 489) calls "peer-to-peer surveillance". As such, sousveillance might pave the way for visual investigation, denunciation and social control on a large scale and to the establishment of new visual/political hierarchies (which conspicuously resemble the established hierarchies). For example, "if someone with a cameraphone snaps you going to a sensitive business meeting and then emails it to a competitor, it is hard to imagine what could [be] done about it".[20] Dennis (2008, pp. 350-352) even reports on cases of "virtual-vigilantism". Finally, practices of spying in connection with sousveillance also include cases of spying on oneself, that is to say, practices of meticulously documenting one's own movements using camera phones, webcams, web-blogs and other such devices. As Dennis (2008, p. 355) notes, there "is a fine line [...] between this being a willing step or one forced upon the individual as an enactment of resistance to hierarchical forms of monitoring and surveillance".

Similar to other forms of photography in the digital age, one of the main challenges concerning sousveillance images is neither the production nor the dissemination of such images but the response to them; the problem is not the collection of data but the transformation of data into emancipatory politics for the sake of global justice or social change, for example. Here, it is useful to think about sousveillance in connection with both the notion of deterritorialized visual/political space suggested by Azoulay (2008) in her discussion of the citizenry of

---

[20]   "The age of sousveillance," *Guardian*, (14/07/2005). Available online: from http://www. guardian.co.uk/technology/2005/jul/14/comment.comment [December 12, 2009].

photography and with the deindividualizing elements underlying Hariman and Lucaites's approach to common spectatorship (2007). Sousveillance images have to be shared with others, both nationally and internationally, and then translated into an emancipatory and progressive joint political response to the conditions depicted in these images. Without a doubt, the second step is more difficult than the first, but without it, a sousveillance image would just be another image. Even if this translation succeeds, Andrejevic's (2005, p. 494) statement still applies, that rather than reflecting a democratization of monitoring, monitoring practices often copy and strengthen top-down forms of control, which are internalized and adapted to the private sphere. In this manner, the patterns on which contemporary surveillance societies are based are reproduced and strengthened. Indeed, as Mann et al. (2003, p. 347) conclude, ultimately "[u]niversal sur/sousveillance may support the power structures by fostering broad accessibility of monitoring and ubiquitous data collection". At the end of the day, then, there seem to be more reasons for concern than for celebration in connection with modern surveillance societies.

8   Conclusion

To the history of street photography as elaborated by Westerbeck and Meyerowitz (1994) and Scott (2007), a new chapter must be added – a chapter on the citizen-photographer. Equipped with digital cameras and camera phones and producing instant images that can be disseminated without delay, the citizen-photographer contributes to the diversification of our views on the world, to the visibility of marginalized groups of people and to the visual construction of political space beyond and to some extent unaffected by governmental control. Digital technologies also offer new possibilities for exposing people and their policies: as the World Press Cartoon exhibitions regularly show, the use of irony and caricature is a potent political weapon. Sousveillance images can be funny,[21] but they also may help to neutralize surveillance images, especially when they are thought of as deterritorializing and deindividualizing. As Danchev (2009, p. 59) writes, "it is given to artists, not politicians, to make a new world order". It is also given to citizen-photographers to make a new world order.[22]

---

[21]   See, for example, http://www.strictlynophotography.com for pictures that were not allowed to be taken.

[22]   The author would like to thank Rune Saugmann Andersen and the editors and the referees for *Images in Mobile Communication* for their thoughtful comments on earlier drafts of this piece. Many thanks also to Peter Kennard and Cat Phillipps for permission to reproduce their photographic montage work and to Alex Danchev for inviting me to the Workshop on art, war and terror, St Anthony's College, Oxford, November 27, 2009.

References

Andrejevic, M. (2005). The work of watching one another: Lateral surveillance, risk, and governance. *Surveillance & Society, 2*(4), 479-497.

Azoulay, A. (2008). *The Civil Contract of Photography.* New York: Zone Books.

Bal, M. (2007). The pain of images. In M. Reinhardt, H. Edwards and E. Duganne (Eds.), *Beautiful Suffering: Photography and the Traffic in Pain* (93-115). Chicago: The University of Chicago Press/Williamstown: Williams College Museum of Art.

Ball, K. & Murakami Wood, D. (Eds.) (2006). *A Report on the Surveillance Society: For the information commissioner by the Surveillance Studies Network.* Available: http://www.ico.gov.uk/upload/documents/library/data_protection/practical_applicati on/surveillance_society_full_report_2006.pdf [May 5, 2010].

Barthes, R. (2000). *Camera Lucida: Reflections on Photography.* London: Vintage. (Original work published 1980)

Benjamin, W. (1963a). Das Kunstwerk im Zeitalter seiner technischen Reproduzierbarkeit. In W. Benjamin, *Das Kunstwerk im Zeitalter seiner technischen Reproduzierbarkeit: Drei Studien zur Kunstsoziologie* (pp. 7-44). Frankfurt: Suhrkamp. (Original work published 1936)

Benjamin, W. (1963b). Kleine Geschichte der Photographie. In W. Benjamin, *Das Kunstwerk im Zeitalter seiner technischen Reproduzierbarkeit: Drei Studien zur Kunstsoziologie* (pp. 45-64). Frankfurt: Suhrkamp. (Original work published 1931)

Borenstein, J. (2009). Camera phone images: How the London bombings in 2005 shaped the form of news. *Gnovis, 9*(2). Available: http://gnovisjournal.org/journal/camera-phone-images-how-london-bombings-2005-shaped-form-news [May 4, 2010]

Campbell, D. (2003). Cultural governance and pictorial resistance: Reflections on the imaging of war. *Review of International Studies, 29,* Special Issue, 57-73.

Couldry, N. (2000). *Inside culture: Re-imagining the method of cultural studies.* London: Sage.

Danchev, A. (2009). *On art and war and terror.* Edinburgh: Edinburgh University Press.

Dennis, K. (2008). Keeping a close watch – the rise of self-surveillance and the threat of digital exposure. *The Sociological Review, 56*(3), 347-357.

Dibdin, M. (1999). *Blood rain: An Aurelio Zen mystery.* London: Faber and Faber.

Downey, A. (2009). Thresholds of a coming community: Photography and human rights. *Aperture, 194,* 36-43.

Ebony, D. (2006). Botero Abu Ghraib. In F. Botero, *Botero Abu Ghraib* (pp. 5-19). Munich: Prestel.

Eisenman, S. F. (2007). *The Abu Ghraib Effect.* London: Reaktion Books.

Foucault, M. (1994). *The Order of Things: An Archaeology of the Human Sciences.* New York: Vintage Books. (Original work published 1969)

Fruzzetti, L. & Östör, Á. (2007). *Singing Pictures: Art and Performance of Naya's Women.* Lisbon: Museu Nacional de Etnologia.

Gilgen, P. (2003). History after film. In H. U. Gumbrecht and M. Marrinan (Eds.), *Mapping Benjamin: The Work of Art in the Digital Age* (53-62). Stanford: Stanford University Press.

Greenough, S. (2009). *Looking in: Robert Frank's The Americans*. Göttingen: Steidl/The National Gallery of Art.

Grey, J. (2009). *1.3: Images From my Phone*. New York: powerHouse Books.

Hariman, R. & Lucaites, J. L. (2007). *No caption needed: Iconic photographs, public culture, and liberal democracy*. Chicago: The University of Chicago Press.

King, B. (2003). Über die Arbeit des Erinnerns: Die Suche nach dem perfekten Moment. In H. Wolf (Ed.), *Diskurse der Fotografie: Fotokritik am Ende des fotografischen Zeitalters* (pp. 173-214). Frankfurt: Suhrkamp. (Original work published 1993)

Koskela, H. (2004). Webcams, TV shows and mobile phones: Empowering exhibitionism. *Surveillance & Society, 2*(2/3), 199-215.

Little, W. (1973). *The Shorter Oxford English dictionary on historical principles* (Vol. 1, p. 109). Oxford: Oxford University Press.

MacDougall, D. (1998). *Transcultural Cinema*. Princeton: Princeton University Press.

Mann, S., Nolan, J., & Wellman, B. (2003). Sousveillance: Inventing and using wearable computing devices for data collection in surveillance environments. *Surveillance & Society, 1*(3), 331-355.

Mitchell, W. J. T. (1994). *Picture Theory: Essays on Verbal and Visual Representations*. Chicago: The University of Chicago Press.

Möller, F. (2009). The looking/not looking dilemma. *Review of International Studies, 35*(4), 781-794.

Phillips, S. S. (Ed.). (2010). *Exposed: Voyeurism, Surveillance and the Camera*. London: Tate Publishing.

Reichel, P. (1991). *Der schöne Schein des Dritten Reiches: Faszination und Gewalt des Faschismus*. Munich: Carl Hanser Verlag.

Ritchin, F. (2009). *After Photography*. New York: W.W. Norton.

Scott, C. (2007). *Street Photography: From Atget to Cartier-Bresson*. London: I.B. Tauris.

Shapiro, M. J. (2009). *Cinematic geopolitics*. New York: Routledge.

Simmonds, J. (2009, November 27). The artist as witness. Presentation at the Workshop on art, war and terror. Oxford: St Anthony's College.

Simpson, D. (2006). *9/11: The Culture of Commemoration*. Chicago: The University of Chicago Press.

Sire, A. (Ed.) (2009). *Photographing America: Henri Carter-Bresson/Walker Evans 1929–1947*. London: Thames & Hudson.

Sliwinski, S. (2004). A painful labour: Responsibility and photography. *Visual Studies, 19*(2), 150-161.

Sontag, S. (1977). *On Photography*. London: Penguin.

Strauss, D. L. (2003). *Between the Eyes: Essays on Photography and Politics*. New York: Aperture.

Westerbeck, C. & Meyerowitz, J. (1994). *Bystander: A History of Street Photography*. London: Thames & Hudson.

Part II

Strategies and Tactics
at the Advent of Mobile Images

Revolution in Journalism?

Mobile Devices as a New Means of Publishing

Cornelia Wolf and Ralf Hohlfeld

Introduction

As a means of interpersonal communication, the mobile phone has experienced an unbelievable boom and is now an integral part of everyday life (Ahonen & O'Reilly, 2007; Carey, 2006, p. 116). There are now more than 4 billion mobile phones worldwide (Bitkom, 2009). In Germany, market coverage stands at more than 130 percent (Bundesnetzagentur, 2009). The reason the mobile phone has been embraced to such an extent can mainly be attributed to social change, which is characterized by the buzzwords "mobility" and "individualization" (Matsuda, 2009, p. 19). Being reachable anywhere and at any time is concomitant with the increasing levels of mobility in society, on both private and professional levels. The convergence of mobility and communication has created the image of an individual who is able to communicate with anyone, anywhere, and in any situation. This total "reachability" is a state that most people have already attained (Hanekop & Wittke, 2005, p. 113). This most innate characteristic of the mobile phone can therefore be considered to be in the personal sphere. All in all, we can speak of "communicative mobility" (Hepp, 2006, pp. 15-21) in a "mobile promised land characterized by ubiquity, connectivity and convenience" (Aguando & Martinez, 2008, p. 69).

Limiting mobile communication to a mere notion of interpersonal, computer-assisted communication would, however, be too much of a simplification. Although mobile devices are still used primarily for one-to-one communication (telephony, SMS; cf. Accenture, 2009, p. 8), the birth of the third-generation mobile device has made available a wide range of new applications that also enable one-to-many communication (Goggin & Hjorth, 2009, p. 3). Technological development is a prerequisite for this transformation of the mobile device into a mass medium[1]; According to Ahonen and O'Reilly (2003, p. 3), "[W]e see

---

[1] Within the sense of Maletzke's classic definition of mass communication, cf. among others Burkart, 2002, p. 171.

convergence of media, web and communications". In principle, a synthesis is taking place between the simultaneous revolution of two technologies in the field of communication, as mobile communication and the internet move ever closer together at a rapid pace (Glotz, Bertschi, & Locke, 2005, p. 12). According to 2009 estimates by the trade association Bitkom, the number of Universal Mobile Telecommunications System (UMTS) connections in Germany rose by 43 percent to approximately 23 million (Bitkom, 2009). "Mobile internet is developing into a mega-trend among consumers," according to Accenture (2009, p. 3). Almost one in five (18 percent) German internet users surfs the web on a mobile phone (ibid., p. 4). As a consequence of digitalization, broadcast-specific, print-specific and online-specific offers are merging with services from the telecommunications field, "resulting in interactive and individualized multimedia offers" (Eimeren & Frees, 2006, p. 563). Nine percent of users receive TV content, and 12 percent download videos or watch them on portals (Accenture, 2009, p. 12). The consensus is that editorially created applications that reach beyond the standard of voice telephony (first generation) and SMS (second generation), and that create a mass communication medium from an interpersonal communication medium (third generation), will determine the success of the new mobile broadcast generation.

On the other side of the equation stand traditional forms of media (such as television and, still to a large extent, the internet), which are either spatially fixed — and, therefore, immobile — or which (in the case of local newspapers) have always been mobile but are limited to a very constricted coverage area (Kretzschmar, 2008). Technological change is now enabling media companies to react to social changes (Wolf, 2008).

"Made-for-mobile" — this magic formula of journalistic-editorial content for mobile devices could lead the way to a form of mobile journalism. From a perspective of diffusion theory, the so-called third generation of mobile phones will achieve a sufficient level of market penetration only if they are able to receive a multitude of specially tailored information and entertainment services (Schweiger, 2002, p. 161). Regarding editorial information services, made-for-mobile formats that shift away from the 1:1 models of traditional content and make use of the specifics of mobile broadcasting are more likely to succeed (Grigoriva, 2007, p. 20f.; Pavlik & McIntosh, 2006, p. 91). Although such applications already exist to a small extent (Breunig, 2006, pp. 550-562), the awareness of them among potential users is either too low, or their trialability cannot be warranted. A systematic review of the production and supply strategies of established media companies in relation to mobile multimedia services (MMS)[2] is not presently available.

Based on two empirical studies from 2007 and 2009, it is possible to establish whether we may already speak of the emergence of a form of mobile journalism. On the one hand, the central objective of the first study was to create an appraisal of the existing and planned activities of media companies, demonstrating how technological potential is used to transmit journalistic information and entertainment. The study also aimed to determine whether there is currently a sufficient level of attractive services and content that enhance the relative use of mobile multimedia services to the extent that the innovation can "take off" in the foreseeable future. On the other hand, the objective was to answer the question of whether and how journalism (in terms of both content and organization) already matches the specifics of the new medium, i.e., the extent to which non-localized communication affects the production of mobile editorial content and the significance that journalistic content has in the field of mobile multimedia services. For this purpose, the TV broadcasters' made-for-mobile news formats were analyzed in the framework of a quantitative content analysis.

Evidence for the institutionalized dispersion of mobile multimedia services is provided by the results of the editorial survey. The survey period was June 15 to July 16, 2007. Although the influence of mobile publishing on journalism was being examined, the study did not take the whole field of journalism into account. Rather, it concentrated on the core area of topical information dissemination. In the sense of a purposive sample, the population was determined based on the selected area of news coverage. It mainly encompassed conventional forms of media that are suited to the publication of mobile information services. For the area of electronic media, this includes TV and radio. For print media, it includes daily and weekly newspapers, as well as news magazines. In addition, news and program suppliers were included in the population for all selected media types; these encompass both news agencies and production companies, including independent third-party suppliers. The Internet was consciously not classed as an independent medium, because it would usually be integrated within the established media companies.[3] The findings can, therefore, not be applied to the entire

---

[2]   Mobile multimedia services include all applications and services that can be received or viewed on a mobile device, that serve the information and entertainment needs of the user, and that are either editorially created by a media company or in its name or distributed under its brand. This may occur either by means of mobile broadcasting or via traditional broadcasting standards.

[3]   Pure online providers are not considered in the study, because the effects on the editorial organization of conventional media are the main focus. For years, conventional media have displayed firm structures. On the contrary, online providers – like multimedia services – allow themselves to be classed in the category of new media. Furthermore, they have been in exist-

field of editorial journalism; at best, they are representative of professionally disseminated, current information from the field of specialized news journalism and its editorial departments. Sampling was based on a mixed process of full sampling and random sampling. A full sample was possible for daily and weekly newspapers, as well as for public and private television and radio broadcasters, both national and statewide. A systematic random sample was used only for private television and radio stations operating locally and regionally. A total of 285 questionnaires were sent out, of which 137 questionnaires were completed (representing 147 sample units), which constituted a satisfactory return of 48.1 percent.[4] In comparing the distribution according to subsamples and returns, it is clear that there is no distortion due to above-average number of returns in one of the subsamples. However, newspapers are represented by a somewhat higher value. Therefore, it can be assumed that the sample validity is high, which allows the sample to be taken as representative.

Many Media Companies Are Already Using the Mobile Channel

As a result of the increasing technological and organizational convergence, the mobile channel is being widely used by German media companies. In 2007, independent multimedia services were being offered by 44 percent of media companies. A further 39 percent were planning to implement them at the time. Only one in six editorial departments did not wish to incorporate MMS in the future.

Accordingly, mobile devices are already being perceived as a new channel by German broadcast and print editorial departments. The numerically largest group of the newspaper editorial departments lies approximately in the middle of the examined sample. Only one in ten newspapers is not considering MMS at present. Ninety percent of newspapers are, to an equal extent, either already in the market or planning to be in the future. Categorized by reach, it is apparent, however, that the proportion of MMS providers among regional newspapers (37%) is much lower than that of national newspapers (80%). However, more than half of the regional newspapers are planning to offer mobile services in the future. Mobile news and entertainment services are even more dispersed in the case of TV broadcasters. More than half the TV editorial departments are already offering MMS. A further 29 percent are in the planning stage. Here, the services

---

ence for only a few years, so one can assume that structures are oriented toward computer-mediated communication from the outset and that technical know-how is more dispersed than it is for conventional media.

[4]   In relation to the sample units, the return stood at 49.3 percent (147/298).

were provided primarily by the national broadcasters. Whereas 83 percent of the national TV broadcasters offer mobile multimedia services, the figure for the regional broadcasters stands at less than 29 percent. The content, broadcast mainly over UMTS, is spread across free-of-charge, complete broadcasts (e.g., Eurosport, CNN Mobile, and n-tv), selected pay-per-view highlights from private broadcasters (e.g., RTL mobile TV, ProSieben mobile, SAT.1 mobile), and free-to-air, public made-for-mobile formats (e.g., "Tagesschau in 100 Sekunden", "Rundschau Handy News"). These are examined in more detail below. Among the radio editorial departments, only 17 percent claimed not to be considering MMS. After the introduction of digital radio in Germany, which experienced only a limited level of success, the broadcasters have evidently realized that mobile devices offer new opportunities to establish themselves in the digital market and to fight the audio competition of music downloads and MP3 players with their own services.

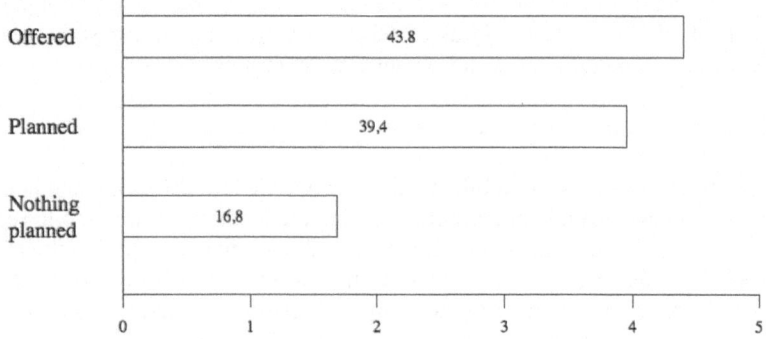

Figure 1: Implementation of mobile multimedia services, N = 137, values in percentages.

Economic reasons are a particular deterrent for entering the market

The main arguments against market entry, as specified by non-providers, include low demand (22%) and waiting to see how the market will develop (22%). In addition, high costs (18%) and the absence of a clearly defined target group (16%) are also important factors in deciding against the publication of mobile services. Less common reasons were a lack of technical possibilities (11%) and insufficient technical know-how (5%). The arguments against the mobile channel are thus much more related to economic factors than to technological ones. On the contrary, there are numerous arguments for entering the market. Both MMS providers and MMS planners primarily aim to address young target groups. This is promising insofar as studies have shown that young people in particular display a higher interest in mobile services (Kaumanns & Siegenheim, 2006, p. 502), while at the same time being the group most equipped with the required devices[5] (Medienpädagogischer Forschungsverbund Südwest, 2009, p. 9). The intention to serve customers through all channels within the framework of a cross-media strategy, and to incorporate MMS as an additional channel into the service portfolio, is equal among MMS providers and planners. The early providers of MMS evidently want to adopt a pioneering role in the market; half of them specified "new technological possibilities" and "image reasons" as motivations for entering the market. To act in a journalistically innovative manner thus plays an important role. For planned mobile activities, it is also important, however, to offer fresher content and to reuse it for mobile multimedia services alongside existing online editing. Among the newspaper editorial departments, it is noticeable that almost two-thirds of MMS providers and almost 80 percent of MMS planners would like to acquire young target groups.

The Right Information, at the Right Time, in the Right Place

Two prominent features of mobile communication lie in localization and context sensitivity. These features potentially combine the location of the mobile reception with the activities that are possible there, taking into consideration the user's preferences. This scenario ensures the right information will always reach the recipient at the right time and in the right place (Buse & Fiedler, 2008). To this extent, the future of mobile information services clearly lies in focusing locally. The surveyed editorial departments mainly take these specific characteristics of multimedia services into account. However, substantial differences exist between MMS providers and MMS planners. World news (including Germany) and sports news are most frequently included under information services that are already available in the market. Local news comes a distant second, followed by soccer results and business news. For planned information services, a clear shift can be seen toward location-based services: Local news and local event tips belong to the preferred type of content to be introduced to the market in the future. As a result, a significant shift away from world news and business news can be expected, and only a fraction of editorial departments in the planning stage specified these as content for future mobile information services.

---

[5]   For 12- to 19-year-olds, this stands at 100 percent.

Future providers are thus focusing on information specific to target groups on at least a local level. In contrast, sports and football news appear to be somewhat change-proof. As expected, the largest activities in the area of mobile information services are performed by newspapers and news magazines because this form of mobile service is the most similar to print media. Furthermore, 57 percent of newspapers are already offering local information services, in contrast to only 8 percent of the TV editorial departments that were surveyed. Nine out of ten newspapers, as opposed to only one-third of TV broadcasters, would even like to give the information service a local orientation in the future. The reason for this is presumably the strong national orientation of the TV broadcasters in the sample. This, however, is counterbalanced by the services in the area of mobile TV, because the described tendency toward a local and regional character is also present here. Although news programs are still being offered most frequently in conventional formats (22%), these programs will become less important in the future (15%). In comparison with the services available today (7%), the regional character of services will develop further in the future (15%), enabling regional mobile TV services to eventually establish themselves equally. Primarily, the described trend will thereby be led naturally by TV providers. However, one in five newspaper editorial departments that are planning MMS indicated that they would also like to set up regional mobile TV. Alongside information services and mobile TV, media companies are also offering mobile entertainment as part of their mobile publications. Economically, mobile multimedia services aimed at entertainment have so far been one of the most successful areas of mobile applications.

Made-for-Mobile Formats are Still a Scarce Commodity

Successful providers of information-oriented online services are known to place emphasis on the exclusivity of content and its media-specific design and preparation. In the field of mobile communication, it can also be assumed that offers for which the specifics of the usage situation have been addressed during their preparation and composition will have a greater chance for success. Sophisticated formats of conventional media, set up as simulcasts, have only limited suitability due to the specific characteristics of mobile devices. The main reasons for this handicap are the small display size of the devices and the application scenarios, which are usually set against a mobile backdrop, during non-occupied time, and often in public places. Across all services (i.e., information services, entertainment services and mobile TV), 39 percent of providers claim to produce contributions or programs specifically for mobile devices. In relation to the possible emergence of a form of mobile journalism, the study shows that only 22 percent of MMS providers claim to produce journalistic content exclusively for MMS. These mobile-specific information services include sports news and regional announcements.

Likewise, the fact that two-thirds of the surveyed editorial departments consider the importance of using mobile content to be low also speaks against the establishment of a form of mobile journalism. Three-quarters indicated that they consider the importance of research for mobile publishing to be low. It is mainly newspaper publishers and TV broadcasters that produce specific made-for-mobile formats. Despite the comparatively high value for mobile-specific formats — based on the information provided by the editorial departments — a rearrangement of the content and forms of conventional media should be the most widely adopted practice at present. Selecting, summarizing and composing existing content and/or its highlights into new topic bundles are clearly the main editorial process (Breunig, 2006, p. 561), whereas the smallest proportion uses content that is identical to that of the mother medium.

News Desk Concepts Benefit Cross-Media Work

Alongside mobile-specific content, an alignment with the appropriate forms of editorial organization and mobile-specific restructuring measures in the media companies would support the development of a form of mobile journalism. The development of news desk concepts that facilitate media-neutral information management, and that also favor the separation of editor and reporter (Meier, 2007 p. 4), can in principle provide good foundations for cross-media — and thus also for mobile — publishing. Editors plan and design, and reporters research and write — across different media. German media companies have already been using this model for several years. The fundamental difference is that, whereas editors previously contested all issues as lone warriors — from identification of topics, to research, implementation and final proofing — the news desk places more emphasis on teamwork beyond departmental boundaries (Meier, 2006; Meier, 2009). Among the surveyed media companies, this form of editorial organization currently stands in second place, and it is practiced by one in three editorial departments.

Overall, almost two-thirds of media companies indicated that an online editorial department exists in-house. Among MMS providers, this is the case in more than four out of five instances, whereas the figure among non-providers of MMS is much lower (one-third). The distribution for planned mobile services corresponds approximately to the overall distribution. Thus, publishing of mobile services at present is closely related to cross-media activities online. If one considers that no online editorial departments were included in the research sample, then there is much to suggest that mobile services are already widely integrated in online cross-media environments (inline), rather than existing only as isolated applications (stand-alone) in connection with conventional media. Thus, the question as to whether the associated online editorial department functions only as a sub-department within the surveyed media companies — or whether it has a separate status with autonomous editorial management — also has an influence on the provision of mobile multimedia services. This is the case for 35 percent of the online editorial departments surveyed. In 54 percent of cases, the online department plays a subordinate role. In the case of the editorial departments that have not yet entered the market with MMS, the number of subordinate online editorial departments lies as high as 66 percent. Only 21 percent are managed autonomously. In comparison, MMS providers are defined by a higher proportion (47 percent) of autonomous online editorial departments; the online editorial department is integrated for only 43 percent of MMS providers. This also suggests a positive connection between established online structure and mobile publishing. To a large extent, the responsibility for executing mobile multimedia services lies with online editors. In 40 percent of cases, mobile publishing is performed by editors from the main editorial department. However, 32 percent of the mobile content is created by online editors, and a further 11 percent is automatically pulled from the internet by editorial systems and content management systems (CMS). If an online editorial department exists at the MMS provider, then the content is primarily generated by the editors in this department.

It would appear that the dispersion of mobile multimedia services is linked to the establishment of online solutions. This finding suggests that the online landscape is becoming the central hub of multimedia channels, to which the mobile channel will also belong in the future. Whether the online-accelerated execution of mobile services will result in a separate form of mobile journalism remains to be seen. From the perspective of journalistic personnel, one would have to say that this is not the case at present: Only five editorial departments currently have editors who are exclusively concerned with preparing mobile multimedia services. These editorial departments, however, do demonstrate some innovation potential, four of them having already taken the leap into the mobile market before 2005.

Conventional Editorial Organization will be Replaced in the Future

Structural changes in the editorial departments can be observed on three levels: Fundamental editorial organization, editorial procedures and editorial personnel. Regarding the procedures, it is proven that the news desk model is assuming increasing importance. Sixty percent of the editorial departments that are willing to restructure are planning to convert to the principle of central information management.

Almost half of the MMS providers have already done so. In the future, this form will thus advance to the preferred model of editorial organization and, in times of the multi-channel strategy, will replace conventional editorial organization. Furthermore, there will be significant changes in general production processes and technology. Around a third of the restructured editorial departments have implemented new editorial and content management systems in the course of pursuing a stronger mobile orientation.

On the contrary, a different picture exists in relation to the fundamental restructuring of the organizational units. When the main question of what makes editorial departments mobile is asked, what is perhaps the decisive factor in the organization is revealed: Almost half of the respective MMS providers have already made radical changes in the form of new departments and editorial departments, as well as creating cross-departmental teams. Around a quarter have formed new departments. The MMS planners who envisage restructuring are significantly more reserved in this regard. Fewer than one in ten editorial departments is planning new divisions or departments. One in three is trying cross-departmental teams. Regarding personnel changes, the tendencies of both groups are the same. Existing employees are trained and educated in mobile publishing to enable them to operate the new distribution channel. As a whole, restructuring by future providers will occur more in the area of procedures, whereas — on a visible level — existing mobile multimedia services were more likely to lead to organizational changes in the editorial departments.

Future Prospects: Journalism 2012

In German newsrooms, there is 96 percent certainty that mobile publishing will become an integral part of journalism. However, the scope and quality of this mobile element of journalism is debatable at present. Alongside computer-mediated communication via stationary and portable computers, it is highly likely that the mobile device is now also responsible for the deep-seated changes in mass communication processes. These changes now not only occur unilaterally, but also are intermittently interactive, enabling constant, direct feedback. According to many of the respondents, one of the most likely editorial consequences of individualization is that news in the future will be researched for target groups and, as a result of possible identification and localization, will even be positioned in the relevant medium in a manner specific to the target group. Within the scope of an open question regarding journalism in the future, the trend was reiterated that every topic will be produced for several channels, namely in the form of a media-specific topicality and relevance chain. The current scenarios for the mobile channel and the internet will be the beginning. The daily newspaper will be seen as the last link in the exploitation chain — but as having the highest quality content. Two effects of the multi-channel journalism related to these scenarios could fundamentally change the editorial organization of German news journalism. First, a greater distinction is expected between pure producers and/or news managers and researching and writing reporters, as is traditional in Anglo-Saxon journalism. Topic specialists will research and write, irrespective of the medium, and channel specialists will prepare the content in a medium-specific manner. Second, cross-media and multimedia processes could soon also result in comparatively fewer, but more specialized and better-paid, editors taking over the production of content and topics, which will be supplied to them by many lower-paid freelance workers for TV, radio, print, online and mobile platforms. This journalistic infantry then will no longer work for the (type of) media, but, rather, for cross-media brands and all their permutations.

The Scope and Quality of a Form of Mobile Journalism Are Still Unclear

The independence of mobile services — and thus the potential for autonomous mobile journalism — should not be overestimated. Mobile multimedia services will in all cases presumably be a complementary, additional channel in the portfolio of media companies operating across different media, which at best will form and strengthen brands. It is likely that mobile multimedia services in online-dominated cross-media environments will profile themselves as "online, small and portable". Positioned in such a way, the eventual best-case scenario is that a niche use is created in this channel for its multimedia services, which should particularly flourish in situations in which people are waiting or traveling. According to this version of future scenarios, mobile services will contribute toward consolidation of multi-channel journalism. Regarding the design of these services, which will soon be offered by a significant majority of German media companies, the future undoubtedly lies in the made-for-mobile principle. Much potential still exists to optimize services for mobile use. In the face of flagging demand, many providers are keeping expensively developed mobile phone–TV formats — i.e., interactive "mobisodes" — under wraps.

Made-for-Mobile Is Put to the Test

For the reasons cited above, mobile TV news formats (which, according to the TV broadcasters, had been developed especially for mobile reception) were specifically examined in a further study in 2009 and compared with their counterparts in conventional television. It was assumed that the mobile information services of current news broadcasts would differ from TV news broadcasts in terms of content and design.

Due to the mobile usage situation (filling short periods of unoccupied time in public places), one can expect a reduction in length, a multitude of design elements and different styles of camera work in the case of made-for-mobile formats. Due to the relatively small display screen of the mobile phone, the presentation type and length of overlays and lower thirds were also examined. For this study, we examined whether the topics and content of TV news are tailored to the usage context of mobile formats, i.e., whether they have a small variety that meets the user's requirement for fast and concise information. In addition to the design and content aspects of mobile information services, we also examined to what extent the current providers produce images specifically for mobile devices — or if, as expected, they merely perform an editorial selection and summarization of existing image material from conventional media.

The examined broadcasts were recorded on July 8, 2009, and analyzed by means of a quantitative content analysis. In total, ten broadcasts were examined for a total of 182 analysis units (contributions). The objects of the analysis were the main news broadcasts, the mobile news formats of the two national public TV broadcasters (ARD and ZDF), and a regional public broadcaster (BR), as well as a private full-content broadcaster (RTL) and a private special-interest broadcaster (N24). Individually, these are "Tagesschau" for ARD compared with "Tagesschau in 100 Sekunden" for mobile TV. For ZDF, the TV program "heute" was compared with "heute in 100 Sekunden", for Bayerischen Rundfunk "BR Rundschau" was compared with the mobile "Rundschau news", for RTL "RTL aktuell" was compared with "RTL kompakt", and for N24 "N24 Nachrichten" was compared with "N24 kompakt". In an additional module, the mobile news formats were studied over the course of the day. These broadcasts were recorded on July 8, 2009, at 10.00, 13.00 and 16.00, respectively and, finally, at the time of the main news broadcast on TV. These broadcasts, too, were analyzed according to the rules of quantitative content analysis.

Formal Differences Exist Between TV and Mobile News

As far as the design of mobile and TV news, it is clear that the portable version is different on both a program and a contribution level. A comparison of the lengths of the analysis units shows clear differences: At 11 seconds, the contributions of the mobile news are significantly shorter than those of the TV broadcasts (37 seconds). At the same time, the number of contributions, as well as the spectrum of presentation formats used, is lower. Whereas TV news exhibits between 14 and 36 contributions, the figure for mobile broadcasts stands at between 7 and 13. Presentation elements[6] are thereby used less frequently in the mobile broadcasts. In all forms, these amount to 33 percent for TV programs and only 21 percent for mobile formats. The greeting makes up the largest proportion (8%) of the presentation for mobile broadcasts. Further evidence of the low relevance of packaging and presentation in mobile broadcasts is demonstrated by the total number of analysis units lacking these two categories. Here, the total number is only slightly different in the case of mobile broadcasts. It is somewhat different in the case of TV formats, for which the number of analysis units is significantly lower. Due to the shorter program duration, linking and complementary presentation elements such as introductory and closing presentation, transitional presentation and presenter discussions are consciously not used in the case of mobile news. In addition to the differences in how presentation elements are used, it is also evident that the diversity is limited for the broadcasts that were examined. Of 23 possible categories, 16 are used in the mobile formats. For TV news, a total of 20 are used. In addition, there are mobile-exclusive design elements such as "read-out news with speaker and theme overlay," "read-out news with graphics in the foreground" and "news in film with theme overlay".

After determining that presentation elements play only a secondary role, the question arises as to whether topics in mobile broadcasts are more likely to be announced by written overlays that offer the recipient some orientation. When considering the percentage proportions of theme overlays in all broadcasts, it becomes apparent, however, that the mobile news broadcasts make only slightly more use of these overlays (37%) than do their TV counterparts (34%). Conversely, this means that more than half of the contributions are sent without lower thirds, i.e., the topic is transported only via the available images and/or voiceover. In general, public broadcasters use theme overlays more frequently than do private broadcasters. As a whole, however, one can say that the opportunity to provide the recipient with written orientation is not used sufficiently. Furthermore, the format "read-out news with background" (presenter before an overlaid image), which is not suitable for mobile usage, is not used. Split screen is also not used.

---

[6]   Presentation elements include introductory presentation, closing presentation, transitional presentation, greeting and exit.

Camera Movement is Restricted

When program and contribution length, as well as the design elements of mobile news, are aligned with the mobile usage situation, how the individual analysis units are arranged in relation to editing technique and camera work becomes interesting. If one now takes the individual analysis units as a basis, the study shows that the number of shots used is significantly lower. Whereas a maximum value of up to 30 shots is reached for the main TV news broadcasts, which last an average of 12 seconds, the number for mobile news formats stands at a maximum of seven shots. On average, the individual contributions consist of around three-and-a-half shots, lasting an average of 7 seconds.

Furthermore, the number of pan shots and camera movements is lower for mobile news than for the TV formats that were studied. In 71 percent of cases, the broadcasters do not use any camera movements within the individual shots for mobile formats. For TV news, the figure stands at 65 percent. Only the number of camera zooms is practically identical for both formats. This finding suggests that the mobile broadcasts, even if only to a small extent, are tailored to the mobile usage situation regarding camera movement. Accordingly, mobile news (with 2 percent) uses more fixed images (photos, diagrams, graphics) in its news than do its corresponding TV formats. In relation to the camera shots, it can be concluded that semi close-up and extreme close-up shots are used more frequently in mobile broadcasts, but that TV formats contain a higher percentage of close-up shots.

Range of Topics in TV and Mobile TV News

Whereas in conventional TV broadcasts, topics from all 11 categories are covered[7], only 4.5 topic areas are covered in the mobile version.

Updating the Programs

How important the currency of the news is for mobile broadcasts is demonstrated by the samples mentioned above spread over a day. Accordingly, one broadcast from all mobile news providers was examined at 10:00, 13:00, 16:00 and 19:00 or 20:00 regarding its coverage. Here it is not only the topic of the report that is important but also where the individual contributions are positioned in the broadcasts. In the majority of mobile formats, the opening topic of the 10:00

---

[7]   The topics were organized in the following categories: National Affairs, Foreign Affairs, Business, Science and Research, Private Lives of Celebrities, Accidents, Human Interest, Environmental Disasters, Sports, Culture, and Weather.

broadcast thus loses importance throughout the day and is completely omitted from the news format in the last time slot. Although this is not the case for every topic in every broadcast, all the broadcasts that were examined demonstrated differences throughout the day, both in the choice of topic and in the positioning of the topic in the schedule. This finding suggests that each broadcast is newly formulated in light of current events.

Regarding the visual material that is used, we found that contributions in which the visual material of the mobile broadcast is completely identical to that of the TV broadcast made up almost half (46%) of all studied analysis units. Twenty percent of the visual material was only partially identical to that used in the TV news version, and 11 percent of the contributions showed visual material that was completely different from that used in the conventional TV format. Around 23 percent of the mobile contributions were not seen in the TV broadcasts (i.e., no visual material was used). Moreover, the differences between public and private TV broadcasters were notable: Whereas the former used more visual material that was not used in the TV broadcasts, the private broadcasters demonstrated a visual material adoption rate of almost 90 percent (i.e., "visual material is completely identical"). Whether "completely different" visual material is produced specifically for use in mobile news, or whether available visual material that has not yet been used for TV is primarily selected and/or composed for mobile formats, cannot be determined from the data. However, it can be expected — especially regarding the low alignment of the design elements — that the production of mobile-specific visual material has not been adopted at this point, and that existing visual material is therefore primarily used.

First Alignments Can Be Seen, But Are Not Yet Sufficient

The main findings of the study support the hypothesis that mobile news formats differ from conventional TV news in their design. As a rule, therefore, the mobile broadcasts are significantly shorter, both on a program and a contribution level, and contain less presentation and linking elements between the individual contributions. In particular, linking and complementary presentation elements such as introductory presentations, transitions and closing presentations are significantly fewer. Furthermore, the presentation elements that occur are shorter in relation to the TV formats. Long presentation formats are avoided, and text-picture combinations are preferred. As far as the shots, there were fewer changes, and the shots did not differ much in length from those of their TV counterparts. An alignment of content with the mobile usage situation takes place only to a limited extent. Individual broadcasters reduce their topic range significantly, however. Based on the study results, it can generally be observed that the potentials for the design of mobile news have not yet been fully recognized or exploited by providers.

In summary, it can thus be said that providers of mobile news broadcasts address only the specific usage situation and the particular technical characteristics of mobile devices, and only to a limited extent. This is reflected in the results of the editorial survey. According to these results, 39 percent of providers claim to produce contributions or programs specifically for mobile devices, but only 22 percent of providers produce specific journalistic content exclusively for mobile devices. This finding corresponds with the research results of the TV study. Until now, the production of mobile formats has been mainly restricted to the rearrangement of content and forms from conventional TV formats. This is attributable to the low rank that the researching and processing functions for mobile content occupy in editorial departments. Due to the lagging demand for mobile phone–TV formats thus far, conventional providers do not yet appear to be ready to invest sufficient time and funds into optimizing their mobile services in line with their specific characteristics related to reception and technical requirements. The success of mobile news programs, however, depends on alignment with the new format. If the user is unable to recognize any added value in consuming mobile news, media companies will not be able to attain a sufficient level of usage to be profitable. The mobile productions by public broadcasters should be seen as the initial benchmarks. In particular, the format used in the program "heute in 100 Sekunden" by ZDF should be noted. Although the program still has shortcomings as far as overlays and camera shots, and in terms of length, presentation and tapering of design elements, it is associated with mobile usage to the greatest degree. As far as content, we need to accept that, even in the future, there will be limited room for depth in the mobile mass medium.

To achieve sufficient added value for the user, it is necessary for mobile news providers to supply content that clearly goes beyond being simply a more compact version of regular TV news; in this way, the exclusivity of mobile news will be ensured. Because the constantly changing geographic location of the user is one of the characteristics specific to the mobile device, it would make sense for the mobile content provider to start with regional information. Thus, we can assume that the trend for the future will "move more towards tailored, context-sensitive local information for mobile recipients" (Hohlfeld & Wolf, 2008, p. 213).

The observation that potentials for the design of mobile news have not yet been fully recognized or exploited by the providers needs to be specified.

Overall, adjustments to mobile specifications in the visual field of mobile news are obviously more pronounced than in the text-based dimension. This result is obvious because the miniaturization of the online channel naturally has a greater effect on the visual design. At the same time, it should be noted that it is harder to adapt visual content to different contexts than it is to adapt textual content. Consequently, the noted changes in the visual material used for mobile communication are the result of pure reduction. Neither the waiver on the production of mobile-specific visual material nor the more economical use of camera angles and camera movements suggest changes in how visual material is applied in mobile communication.

References

Accenture (2009). *Mobile Web Watch 2009.* Available: http://www.accenture.com/ SiteCollectionDocuments/Local_Germany/PDF/Accenture_MobWebWatch2009_ Studie.pdf[May 25, 2011].

Aguando, J. & Martinez I. (2008). The Forth Screen and the Liquid Medium: Notes for a Characterization of the Media Cultures Implicit in Mobile Entertainment Contents. In M. Hartmann, P. Rössler and J. Höflich (Eds.), *After the Mobile Phone? Social Changes and the Developement of Mobile Communication* (pp. 69-84). Berlin: Frank & Timme.

Ahonen, T. & O'Reilly, J. (2007). *Digital Korea. Convergence of Broadband Internet, 3G Cell Phones, Multiplayer Gaming, Digital TV, Virtual Reality, Electronic Cash, Telematics, Robotics, E-Government and the Intelligent Home.* London: Futuretext.

Ahonen, T. & O'Reilly, J. (2007). *Digital Korea: Convergence of Broadband Internet, 3G Cell Phones, Multiplayer Gaming, Digital TV, Virtual Reality, Electronic Cash, Telematics, Robotics, E-Government and the Intelligent Home.* London: Futuretext.

Bitkom (2009). *Mehr als vier Milliarden Mobilfunknutzer weltweit.* Available: http://www.bitkom.org/de/presse/30739_60608.aspx [December 11, 2009].

Breunig, C. (2006). Mobiles Fernsehen in Deutschland: Angebote und Nutzung. *Media Perspektiven, 11,* 550–562.

Bundesnetzagentur (2009). *Teilnehmerentwicklung und Penetration in deutschen Mobilfunknetzen.* Available: http://www.bundesnetzagentur.de/media/archive/16875. pdf [December 11, 2009].

Burkart, R. (2002). *Kommunikationswissenschaft.* Wien, Köln, Weimar: Böhlau Verlag.

Buse, S. & Fiedler, F. (2008). Perspektiven des Mobile Commerce in Deutschland. Erfolgsaussichten Mobiler Informationsdienste. In S. Buse and R. Tiwari (Eds.), *Perspektiven des Mobile Commerce in Deutschland: Grundlagen, Strategien, Kundenakzeptanz, Erfolgsfaktoren* (pp. 289-429). Aachen: Shaker Verlag.

Carey, J. (2006). Contents and Services for Next Generation Wireless Networks. In Groebel, J., Noam, E. and Feldman, V. (Eds.), *Mobile Media. Content and Sercvices for Wireless Communications* (pp. 114-130). Mahwah, London: LEA.

Eimeren, B. & Frees, B. (2006). Zukünftige Medien: Praxistauglich für den Konsumenten? Eine Analyse auf Basis der Daten der ARD/ZDF-Online-Studie und der ARD/ZDF-Studie Massenkommunikation. *Media Perspektiven, 11*, 563–571.

Glotz, P., Bertschi, S. & Locke, C. (2002). *Thumb Culture. The Meaning of Mobile Phones for Society.* Bielefeld: transcript.

Goggin, G. & Hjorth L. (2009). The Question of Mobile Media. In Goggin, G. and Hjorth L. (Eds.), *Mobile Technologies. From Telecommunications to Media* (pp. 3-8). New York, London: Routledge.

Grigorova, P. (2007). *Das Handy der dritten Generation. Symbolmedium einer neuen drahtlosen Gesellschaft.* Saarbrücken: Verlag Dr. Müller.

Hanekop, H. & Wittke, V. (2005). Die Entwicklung neuer Formen mobiler Kommunikation und Mediennutzung. In S. Hagenhoff, D. Hogrefe, E. Mittler, M. Schumann, G. Spindler and V. Wittke (Eds.), *Göttinger Schriften zur Internetforschung* (pp. 109-137). Göttingen: Universitätsverlag Göttingen.

Hepp, A. (2006). Kommunikative Mobilität als Forschungsperspektive: Anmerkungen zur Aneignung mobiler Medien- und Kommunikationstechnologie. *Ästhetik & Kommunikation: Mobil kommunizieren, 135*, 15-21.

Hohlfeld, R. & Wolf, C. (2008). Media to go: Erste Konturen eines mobilen Journalismus? *Media Perspektiven, 4*, 205-214.

Kaumanns, R. & Siegenheim, V. (2006). Handy-TV: Faktoren einer erfolgreichen Markteinführung. Ergebnisse einer repräsentativen Primärstudie. *Media Perspektiven, 10*, 498-509.

Matsuda, M. (2009). Discourses of Keitai in Japan. In M. Ito, D. Okabe and M. Matsuda (Eds.), *Personal, Portable, Pedestrian: Mobile Phones in Japanese Life* (pp. 19-38). Cambridge, London: The MIT press.

Medienpädagogischer Forschungsverbund Südwest (2009). JIM-Studie. Available: http://www.mpfs.de/fileadmin/JIM-pdf09/JIM-Studie2009.pdf [December 11, 2009].

Meier, K. (2006). Newsroom, Newsdesk, crossmediales Arbeiten: Neue Modelle der Redaktionsorganisation uns ihre Auswirkungen auf die journalistische Qualität. In S. Weischenberg, W. Loosen and M. Beuthner (Eds.), *Medien-Qualitäten: Öffentliche Kommunikation zwischen ökonomischem Kalkül und Sozialverantwortung* (pp. 203-222). Konstanz: UvK.

Meier K. (2007). Innovations in Central European Newsrooms: Overview and case study. *Journalism Practice, 1*(1), 4-19.

Meier, K. (2009). Gruppenarbeit in der Zeitungsredaktion. In: C. H. Antoni, E. Eyer and J. Kurscher (Eds.), *Das flexible Unternehmen: Arbeitszeit, Gruppenarbeit, Entgeltsysteme* (pp. 1-30). Düsseldorf: Symposion Publishing.

Pavlik, J. & McIntosh, S. (2006). Mobile News Design and Delivery. In J. Groebel, E. Noam and V. Feldman (Eds.), *Mobile Media: Content and Sercvices for Wireless Communications* (pp. 87-95). Mahwah, London: LEA.

Schweiger, W. (2002). Das hyperaktive Publikum als Dukatenesel? Überlegungen zur Akzeptanz mobiler Mehrwertdienste am Beispiel UMTS. In: M. Karmasin and C. Winter (Eds.), *Mediale Mehrwertdienste und die Zukunft der Kommunikation* (pp. 157-176). Wiesbaden: Westdeutscher Verlag.

Wolf, Cornelia (2008). *Mobile Endgeräte als Allroundmedien: Eine Untersuchung zur Verbreitung mobiler Multimediadienste und ihren Auswirkungen auf den Journalismus.* Saarbrücken: Verlag Dr. Müller.

Which Place for Mobile Television in Everyday Life?
Evidence from a Panel Study

Thilo von Pape and Veronika Karnowski

1   Introduction

Television and mobile telephone constitute two media technologies that are strongly anchored in most consumers' everyday life. Hence, combining those two seems a promising bet. Not surprisingly, early expectations on the success of mobile TV were extremely confident (European Commission 2008, Abiniak, 2008, Cugnini, 2008). While no reliable data for a realistic evaluation of the global market for mobile TV can be found today, there is a general agreement that the early expectations were far too optimistic. As Goggin sums up in 2010 , "consumers have been slow to turn to mobile television" (Goggin, 2010, p. 83).

This situation may partly be attributed to the limits of today's technology – in general the Universal Mobile Telecommunications System (UMTS). Based on point-to-point-technology, it suffers from instability and a loss in image quality once an important number of people use the technology simultaneously in the same area (Gaunt, 2007). The hopes thus draw upon new broadcast-based technologies permitting an unlimited number of users to receive mobile television without any lack in quality. The most widely used standard is Digital Video Broadcasting – Handheld (DVB-H), which is namely used in Europe (Breunig, 2006). An alternative, which is already well established in South Korea, is the Digital Terrestrial Multimedia Diffusion which works on the base of digital radio, but provides a limited offer of approximately six channels (Lee & Kwak, 2005), whereas the DVB-H allows access to more than 30 channels. In the US, Qualcom's MediaFLO is a comparable service competing with DVB-H and the DTV-standard, propagated by the Open mobile Video coalition (OMVC), an alliance of U.S. commercial and public broadcasters formed to accelerate the development and rollout of mobile DTV products and services.

Nevertheless, the experience made with other promising innovations shows that a sophisticated technology does not guarantee a successful social appropriation and success among the users. Thus, the multimedia messages (MMS) have never achieved the same success as text messages (Delanay 2008), and the Wireless Application Protocol (WAP), providing an Internet connection via mobile phone, has mostly failed (Hung, Ku & Chan 2003). As for visual phoning – which had first been launched in the 1920s and put into serial production in the 1960s –, it still remains a marginal technology (Kraut & Fish 1995; Noll 1992), only slowly gaining ground in the context of voice over IP and messaging services. The drawbacks hindering these innovations' success had often been overseen, partially because the economical or political will to see the innovation in question as a success had distracted the actors from a realistic perspective, replacing it either with purely normative arguments or with questionnable empirical research, which was often not uninterested. As an example for the former, Richard Goggin refers to EU Commissioner Viviane Reding's numerous assertions on the future of mobile TV (European Commission, 2007), paraphrasing them as follows: "Mobile television needs to be a success for the econobenefit of Europe, thus consumers need to be made aware – in effect, taught – about its virtues." As for the empirical research on the potential of mobile TV, Schuurman and colleagues (2009, p. 304) already cautioned in 2009 that "much of the existing literature on mobile television is dominated by surveys conducted or financed by telecom, mobile or media players with strategic interests in the results."

These experiences remind of the importance of accompanying the technological progress with an impartial research, starting out from a users' perspective. Analyzing the appropriation of existing offers allows for better evaluating what an innovation may mean for the user.

First, we will present a theoretical framework which permits to identify the appropriation process of mobile telephones on the level of pragmatic use, symbolic signification and communication about the technology, referred to as "meta-communication" (Wirth, von Pape, & Karnowski, 2008). Subsequently, we will show how the questions coming to the fore in this perspective are translated into a combination of qualitative and quantitative methods. In the results section, our findings from both empirical approaches are presented. In conclusion, we will draw some advices for the future development of mobile television.

2   A Circular Model of Mobile Phone Appropriation

Most of the studies on the appropriation of mobile communication services draw upon one of the two following strategies.

- Either they identify and evaluate psychological as well as sociological factors determining the *acceptance, adoption* and *diffusion* of the technologies on the basis of models like the Theory of Planned Behavior (Ajzen, 1985; cf. Bouwman, Carlsson, Molina-Castillo & Walden, 2007; Shin, 2007), the Technology Acceptance Model (Davis, 1989; cf. Pedersen & Nysveen, 2003; Wang, Lo & Fang, 2008), diffusion research (Rogers, 2003; cf. Hsu, Lu & Hsu, 2006) or combinations of these models (Venkatesh, Morris, Davis & Davis, 2003; cf. Wu, Tao, & Yang, 2007; De Marez, Vyncke, Berte, Schuurman & De Moor, 2007).

- Or they aim to understand the interplay of development factors in the everyday life of specific users or user groups in a more qualitative sense, which goes beyond the dichotomy of adoption vs. rejection. This is generally done by studies based on qualitative methods, often including ethno-methodology (O'Hara, Mitchell & Vorbau, 2003; cf. Taylor & Harper, 2003; Höflich & Hartmann, 2006; Hjorth, 2008; Ling, 2004). With the uses-and-gratifications approach, there is also one approach with quantitative methodology dedicated to a more differentiated perspective on new ICT use (Ruggiero, 2000; cf. Leung & Wei, 2000; Peters & Ben Allouch, 2005). Seldom do the studies integrate elements of both perspectives, for instance by considering the evolution of usage in the course of time (Carlsson, Carlsson, Puhakainen, & Walden, 2006).

The approach applied in the presented study integrates elements of both perspectives, by combining four elements (Wirth, Karnowski & von Pape, 2007; Wirth et al., 2008):

- *Usage and handling* of the technology: This question comprises the decision of adopting a new technology (Davis, 1989; Ajzen, 1985; Venkatesh et al., 2003) and the social process of its diffusion (Rogers, 2003), but also its integration into the spatial and temporal context of the users' everyday life, which is particularly analyzed with the aid of the domestication approach in Cultural Studies (Silverstone & Haddon, 1996; Haddon, 2006; Berker, Hartmann, Punie & Ward, 2006).
- The *symbolic* dimension of prestige and social identity: The question of how the users use the technology in the capacity of a prestige object in order to define their social identity is mentioned by the domestication approach (Silverstone & Haddon, 1996; Haddon, 2006), but also by the Uses-and-Gratifications approach in communication studies (Leung & Wei, 2000; Peters & Ben Allouch, 2005; Wei, 2008).
- The *meta-communication* about mobile communication, i.e. the way users negotiate among each other the norms of usage as well as its social significations (Wirth et al. 2008). This question is particularly treated by the ap-

proaches about the social construction of technologies (Bijker, Hughes & Pinch, 1987; Flichy, 1995; Latour, 2005; Williams & Edge, 1996), but also by frame analysis (Goffman, 1974; Ling, 2004).

- The *time* that goes by in the course of the appropriation process. Several of the mentioned approaches have developed concepts to describe this process, namely domestication (Silverstone & Haddon, 1996; Haddon, 2006; Lehtonen, 2003), social construction of technology (Bijker et al., 1987; Flichy, 1995; Latour, 2005; Williams & Edge, 1996), Uses-and-Gratifications (Peters & Ben Allouch, 2005) and diffusion research (Rogers, 2003).

The continuous evolution of pragmatic usages and their symbolic signification under the influence of meta-communication is described in a circular model: The appropriation is ne-gotiated and renegotiated by the users in a constant process (Wirth et al., 2007) (figure 1):

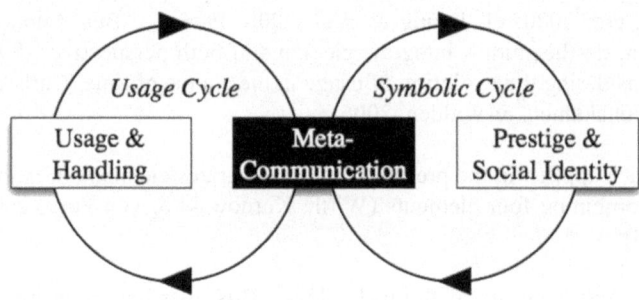

Figure 1: Circular model of mobile phone appropriation (Wirth et al., 2007)

During the process of appropriation, usage and handling as well as prestige and social identity are constantly developing and changing. In time, habitual usage forms emerge and stabilize as well as social evaluations of the symbolic value of certain usage forms, namely their appropriateness or style.

2.1 Research Questions

With the above introduced theoretical model, we can decline the overall research question on the place of mobile television in user's everyday life to four specific questions:

1. Usage: In which situations and for which motives do participants use mobile television?
2. Prestige: Which symbolic value do they associate with it? How does watching mobile television affect their personal and social identity?
3. Meta-communication: How do the participants communicate about mobile television among each other and with non-users? How does their environment react?
4. The evolution of appropriation: How do these factors evolve over time?

2.2 Method

The longitudinal study was realized between September and December 2006. During this period of time, 44 persons received a third generation (3G) phone (model Nokia 6280, figure 2) with a SIM card of the access provider Vodafone.

In order to analyze all the elements of appropriation that had been evoked by the circular model of appropriation, two complementary methods were applied:

• The usage was observed by means of the *Experience Sampling Method*. This method had been developed in sociological research about the everyday life (Larson & Csikszentmihalyi, 1983). It has hardly been applied to analyze the appropriation of the mobile telephone (Palen, Salzmann & Youngs, 2000), but it constitutes the foundation of the presented study.
• Information about the symbolic value and meta-communication was collected in the *guided interviews* realized by telephone.

2.3 Sample

In order to imitate a future in which mobile television will be more widespread, five networks of friends were chosen as participants. By recruiting entire groups of friends, we permitted our participants to exchange their experiences with the technology with each other, as would be the case in a situation of advanced diffusion. The choice of networks targeted innovators, who are, according to the theories of diffusion, rather technophile and wealthy (Rogers, 2003) – the latter criterion also being demanded by our partner Vodafone in order to come close to their principal target group.

The first network comprises ten pupils of a high school in the suburb of the city of Munich, Germany, who were at the average age of 18.5 years. The second network consisted of ten students from the city of Jena, with the average age of 22.8 years. The three other networks comprise employees of a high socio-professional level (net revenues of more than 2,500 Euros [corresponding to about 3.200,- US Dollars in the end of 2006]. They are colleagues who work in the same company and see each other frequently in their leisure time. Overall, these groups consist of 22 men and 2 women at the average age of 36.0 years.

Proceeding

All participants had free access to the complete offer proposed by Vodafone live! for their 3G clients. This offer comprises seven basic channels including Euro-sport, CNN and the German news channel n-tv as well as diverse entertainment channels that are normally part of a superior flat rate – such as MTV mobile3, Bundesliga live (live Premier League soccer games), Pro Sieben mobile (television series) and Nick (program for children). In order to best anticipate the future technological context of mobile television, the participants also had free access to the Internet, MP3 downloading and visual phoning.

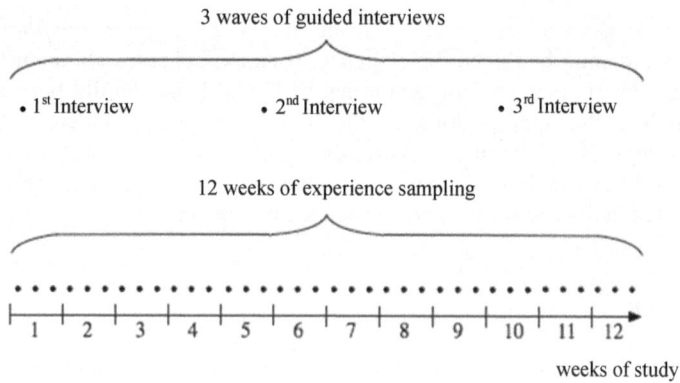

Figure 2:   Proceeding of data gathering in the course of the study

The findings were collected on the occasion of three guided interview waves by telephone (around one, six and twelve weeks after the beginning of the survey) and during twelve weeks by means of a daily sampling of experiences (figure 2).

Data Gathering

In order to collect a representative sample of experiences, every participant was contacted every day by a text message at a contingent moment between 8.00 am and 10.00 pm. The text message comprised the link to an online questionnaire, which was accessible via the Wireless Application Protocol and was opened in the web navigator integrated in the telephone. The questionnaire had to be completed within two hours – after this time it was not accessible anymore.

Thus, the users were interrogated about the services they had been using during the last two hours. If they had used mobile television, they were asked for how long and in what kind of situation they did so (for instance, at home, on public transport, etc.). Likewise, questions about the used programs and their satisfaction with the offer were asked.

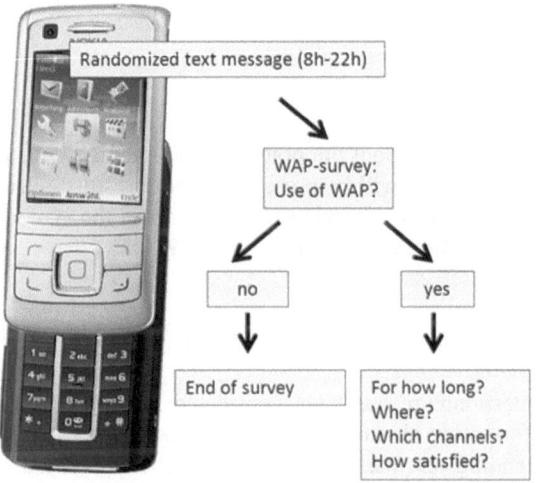

Figure 3: Structure of mobile phone-based survey

The interviews were realized by telephone and comprised approximately ten open questions: how did the participants feel when they used the devices in presence of their families and friends as well as in public? How did they talk about the technology among themselves and with others? etc.

Methodological Caveats

Women are underrepresented in the sample. This is partially due to our difficulty of finding women earning more than 2,500 Euro net and willing to participate in the study with the obligations given by our methodology (interviews, daily surveys).

Another limit consists in the realization of the interviews by telephone, which often makes the respondents express themselves more restrained than they would do in a face-to-face situation.

The decision to offer all the 3G telephone services free of charge to the participants might seem unrealistic taking into account that the providers will eventually want to refinance their services. However, excluding the economic factor allowed us to focus entirely on the often neglected social factors of usage. As for the very advantageous economic conditions of our study, we took them into account in our interpretation of the results.

3   Results

First of all, the findings about the practical use – gained by applying the Experience Sampling Method - will be presented. These results will then be completed by a partial analysis of the interviews, namely about the symbolic dimension and the meta-communication concerning mobile TV.

Use of the Mobile Telephone and of Mobile TV

The online questionnaire has been completed 2444 times in total. Hence, we dispose of information about 2444 utilization periods of two hours length each. Calling and sending text messages are the most frequently used functionalities, whereas mobile television does not show more than 155 utilizations, which corresponds to 6.3 % (see figure 4).

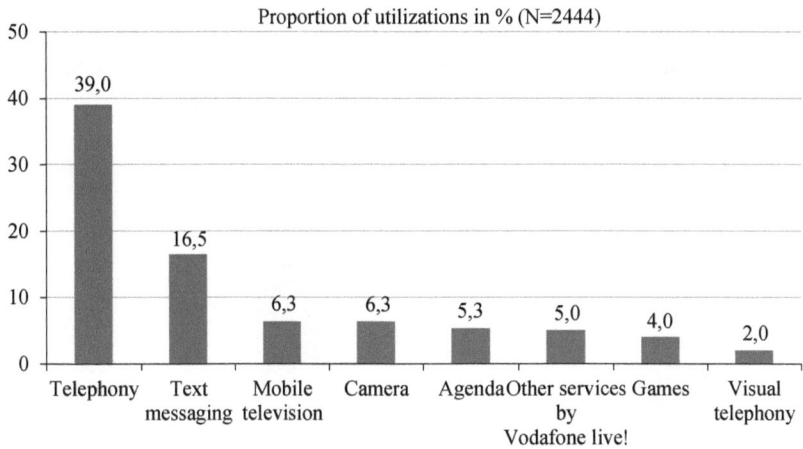

Figure 4:   "Which functionality did you use during the last two hours?"

Almost three quarters of the 155 utilization sessions of mobile television took less than 15 minutes. Hence, its usage is rather seldom and quick (figure 5).

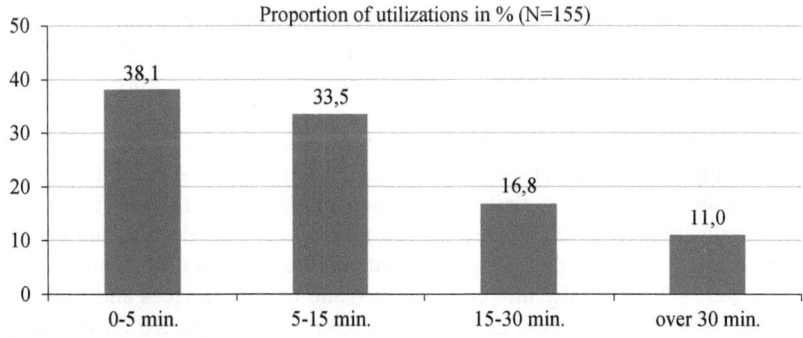

Figure 5:   "During how much time did you use mobile television in the course of the last two hours?"

In the course of the day, the main period of utilization is between 6 pm and mid-night (more than 50 % of sessions). This finding corresponds with the place of utilization because – in half of the cases – the spatial setting is at home and not in public (figure 6).

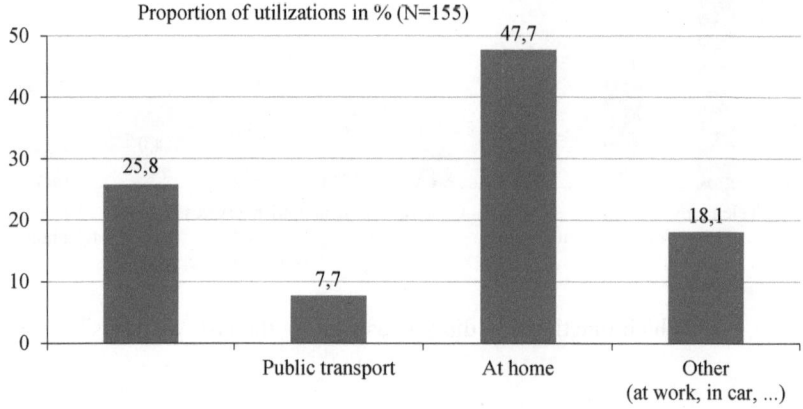

Figure 6: "In which situation did you make use of mobile television?"

It remains to examine how the number of mobile television utilizations develop-ed. After a first period of playful exploration (phase I), it diminishes and stabi-lizes at a low level from the fourth week forward (phase II, cf. figure 7).

This stabilization of usage is crucial for the success of a service. Therefore, we will study the second period of the survey (when the total of used functionali-ties diminishes) in detail. A first look on the overall uses of the different mobile phone applications provided shows an overall decline. This decline can partially be explained by the fact that the participants had received a new device for free use and were thus naturally incited to explore and test all services and functiona-lities available at the beginning. Hence, one could have expected an overall de-cline in use. However, this decline is particularly rigorous for the mobile televi-sion, which passes from 10.7 % of the utilizations in the first period to 3.3 % in the second period. Thus, it did not at all succeed in establishing its position in the everyday life of the participants – even though they could use it for free (cf. figure 8).

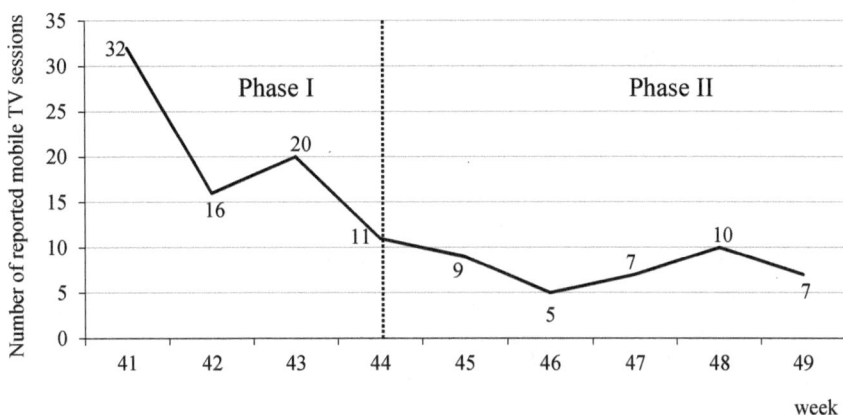

Figure 7: Mobile Television: number of sessions per week (N=155; beginning and end of the survey vary between the groups; we only present the nine weeks when everyone participated simultaneously)

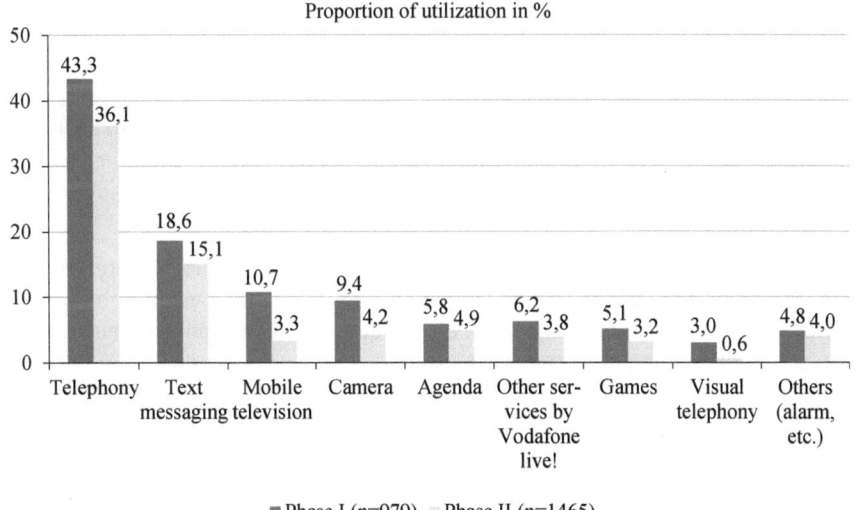

Figure 8: "Which functionality did you use during the last two hours?"

Despite the disappointing findings, it may at least be pointed out that the single utilization periods are longer during the second phase than in the course of the first phase. This shows explicitly that the utilization is not anymore a matter of playful exploration of the service, but a real watching of the programs (cf. figure 9).

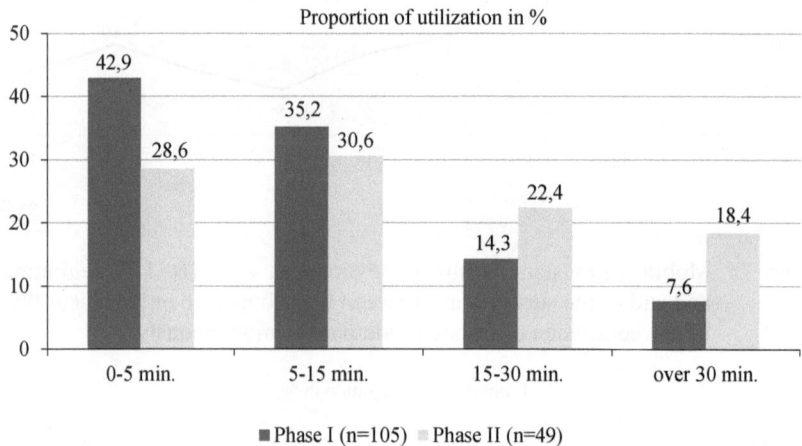

■ Phase I (n=105) ▪ Phase II (n=49)

Figure 9: "During how much time did you use mobile television in the course of the last two hours?"

The domestic usage still prevails in the second period (cf. figure 10). Hence, mobile television remains a service competing with classic television at home and does not succeed in filling any usage niches in moments of transition and mobility (waiting time, on public transport, etc.).

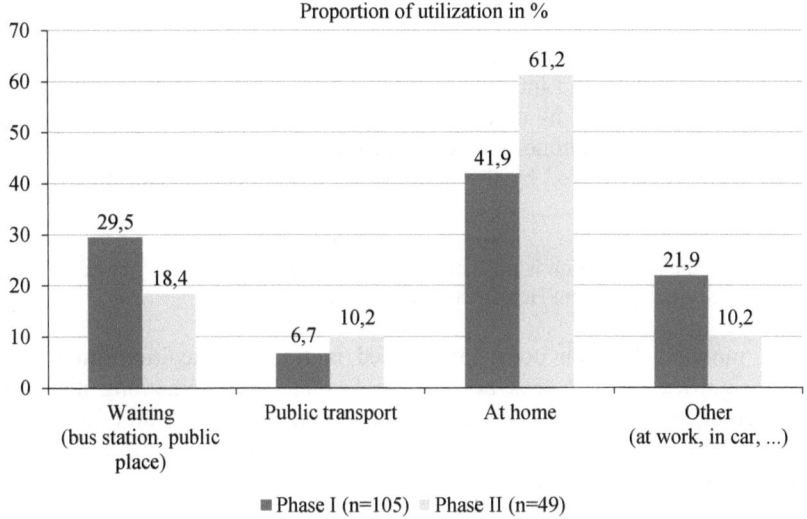

Figure 10: "In which situations have you been using mobile television?"

This conclusion was confirmed by the statements of the participants in the course of the interviews:

> "The thing is, that, let's say, the occasions for usage were not really given." (Peter, 33, employee from Heidelberg, interview 3)

The only single usage pattern which could clearly be identified both on the basis of quantitative and qualitative findings was that a considerable number of participants watched the Premier League soccer matches, which were offered live namely on early Saturday evenings. These uses were mainly at home.

An analysis of the qualitative interviews will finally give an idea of the evolution on the symbolic dimension as well as on the level of meta-communication.

Symbolic Dimensions

At the beginning, the participants' curiosity was aroused by the new service, and they showed an open and interested attitude. This positive attitude is accompanied, as aforementioned, by an exploratory and playful usage of the service and the entirety of the telephone. However, this enthusiasm about the novelty gets lost by and by, as evinced by the participants in the second and third wave of interviews:

> "I think the main reason is that somehow the interest got lost, the 'newness'." (Herbert, 28, employee from Heidelberg, interview 3)

Hence, mobile television does not succeed in finding a legitimization and a proper signification that goes beyond the fact of being something new. This characteristic rapidly loses its pertinence, accelerating the downfall of the appeal. These two phenomena are capable of provoking the decrease of utilization in the second period of the survey.

Meta-Communication

The communication about the new technology presents itself in a quite analogous manner to what has been stated about the symbolic value: At first, the participants often talk about it with their friends and acquaintances – who are equally interested in the technology and its usages. People benefit from showing the functioning to their personal environment.

> [I talk about it] with friends, but also with the family, and with acquaintances, that is to say the neighbors etc. – and of course with the colleagues that have the same device. There, we talk a lot about this subject. After all, with everyone that is around oneself." (Michael, 42, employee from Munich, interview 3).

However, this initial enthusiasm calms down quickly.

> "It was not during the last four weeks, it was at the beginning that we talked about it." (Herbert, 28, employee from Heidelberg, interview 2)

In meta-communication, it can also be observed that mobile television does not keep its positive image. In the course of the discussions, it is more and more regarded as a superfluous gadget:

"It is generally considered a gadget. Like nothing of importance." (Oliver, 43, employee from Munich, interview 2)

In the second period, the – very limited – meta-communication deals mainly with the technical deficiencies of the services. Thus, the participants criticize the heavy electricity consumption and the limited run-time of the batteries. Furthermore, the long connection delays, the frequent interruptions of the connection as well as the complex navigation necessary for watching a mobile television program are points of criticism.

Except for these remarks, the participants rarely speak about the programs provided via mobile television. It thus fails to integrate into the participants' daily communication.

4 Conclusion

The presented study aims to stimulate a discussion about the uses of mobile television at the moment where – thanks to the technical progress – the breakthrough of this technology is expected.

While the technical characteristics of the new technology – and of the latest evolutions – are well-known, the social factors are much more difficult to anticipate because they arise in the course of the process of diffusion and appropriation of the innovation. Hence, the theoretical and methodological challenge is to anticipate the social evolution of this technology.

The circular appropriation model served as a framework to analyze this evolution and to integrate approaches from different traditions. Methodologically, we relied on a longitudinal design, to combine qualitative interviews and experience sampling and to establish our sample on the basis of small groups of friends instead of isolated individuals.

Despite the ideal economic conditions – free of charge services for all the participants – mobile television did not succeed in getting established in their everyday life. The interest in the technology got lost after a short period of exploration, presentation and exchange, which was provoked by the attractiveness of the novelty and possibly enhanced by the thrill participants in any field study feel when asked to evaluate a novelty ("Hawthorne effect", Parsons, 1974). The use decreases, indicating that the technology does not find its place in the everyday life of the users. The only niche identified was for watching free first League Soccer matches live, in the early evening and at home. At the same time, the prestige value diminishes and the technology is finally considered a simple gadget, which is rather ignored in the daily communication.

So, where in the users' everyday life does the potential of mobile television lie?

Probably, this potential does not reside in the usage forms as they exist today: even a mobile phone that works much better could not prevail at the place where and at the time when participants used the technology the most often, that is to say at home in the evening. In this context, the technology competes with domestic television, which already provides impressive technical advances like huge screens and Dolby surround sound systems. The only true advantage of mobile TV to these alternatives in the domestic environment is that it permits to be mobile at home, and hence to move within and between the socially and technologically defined "mediatopes" of the domestic space (Quandt & von Pape, 2010). If anything, then tablet computers may be the ideal compromise for this context, combining a limited mobility sufficient for the home with a limited quality of usage sufficient for some content.

However, this does not answer the question how mobile TV could be used on mobile phones. Additional ideas and efforts are needed to turn this promising technology into an actual success. On the basis of our findings, we see two directions, towards which the evolution of mobile television could be directed:

- Presenting a mobile television of sufficient quality, which can be used on the move and in public spaces, as, for instance, on public transport. First and foremost, this needs excessive labor on the transmission standards, the devices and contents. The arrival of new broadcast standards with improved quality is evidently a first important step in this effort, but it will need to be accompanied by other technological progresses (namely longer-lasting batteries), as well as efforts on ergonomics and contents.
- Giving to mobile television a completely different social signification than just the one of 'traditional television on the move'. One strategy to attain this goal could be to position mobile television in the tradition of an mp3 player. This solution corresponds somewhat to the evolution of Apple's iPod, which turned from a highly prestigious mp3-player into a player of audiovisual material and into a mobile telephone, and which could soon also integrate reception of programs broadcast via DVB-H or other formats. An alternative strategy could be to inscribe mobile TV into the tradition of a creative exchange of videos, which is very common among adolescents nowadays. This idea complies, for example, with the concept of 'grab and share' of audiovisual material on mobile devices, which has been developed and promoted by Microsoft (Harper, Regan, Rouncefield, Rubens & Al Mosawi, 2007): according to this idea, audiovisual content should be presented to consumers as a resource, permitting them to pick out elements, to tag and

potentially modify them, and to share them via networking technology with their peers. In terms of the presented appropriation model, these measures would ask for allocating a signification and a place in the daily communication to mobile television – which goes beyond the simple fact that the innovation is new. There is not only a need for rethinking the uses, but also the symbolic perception, that is to say its value and prestige, and the meta-communication about the service. This is a task that implies the design and the marketing just as the technology itself.

References

Ajzen, I. (1985). From intentions to actions: A theory of planned behavior. In J. Kuhl & J. Beckman (Eds.), *Action-control: From cognition to behavior* (pp. 11-39). Heidelberg: Springer.

Albiniak, P. (2008). What Networks want: Top network executives will roam the halls to plot their technological future. *Broadcast & Cable*, 03/17/2008, 22.

Berker, T., Hartmann, M., Punie, Y., & Ward, K. J. (Eds.) (2006). *Domestication of Media and Technology*. New York: McGraw-Hill.

Biddiscombe, R. (2008). Still waiting for TV-to-go. *Television, 45*(4), 22-23.

Bijker, W. E., Hughes, T. P., & Pinch, T. J. (Eds.) (1987). The Social Construction of Technological Systems: New Directions in the Sociology and History of Technology. Cambridge: MIT Press.

Bouwman, H., Carlsson, C., Molina-Castillo, F. J., & Walden, P. (2007). Barriers and drivers in the adoption of current and future mobile services in Finland. *Telematics and Informatics, 24*(2), 145-160.

Breunig, C. (2006). Mobiles Fernsehen in Deutschland. *Media Perspektiven, 9*(1), 550-562.

Carlsson, C., Carlsson, J., Puhakainen, J., & Walden, P. (2006). *Nice Mobile Services do not Fly. Observations of Mobile Services and the Finnish Consumers*. Paper presented at the 19th Bled eConference. Bled, Slovenia.

Cugnini, A. (2008). Video for mobiles: The industry needs to act fast to take advantage of this opportunity. *Broadcast Engineering, 50*(6), 22-28.

Davis, F. D. (1989). Perceived Usefulness, Perceived Ease of Use, and User Acceptance of Information Technology. *MIS Quarterly, 13*(3), 319-340.

Delanay, J. (2008). MMS five years on. *Journal of Telecommunications Management, 1*(1), 69-78.

Dickson, G. (2008). Qualcom plots use for new spectrum. *Broadcasting & Cable*, 04/07/2008, 23.

European Commission (2007). *Commission opens Europe's Single Market for Mobile TV services*. Available online: http://europa.eu/rapid/pressReleasesAction.do?reference =IP/07/1118&format=HTML& [May 26, 2011].

European Commission (2008). *Mobile TV across Europe: Commission endorses addition of DVB-H to EU List of Official Standards.* Available: http://europa.eu/rapid/press ReleasesAction.do?reference=IP/08/451 [May 25, 2011].

Flichy, P. (1995). *L'innovation technique: Récents développements en sciences sociales vers une nouvelle théorie de l'innovation.* Paris: La Découverte.

Gaunt, J. (2007). *MobileTV and Video: Big Dreams for the Smallest Screen.* New York: eMarketer.

Goffman, E. (1974). *Frame Analysis: An Essay on the Organization of Experience.* New York: Harper and Row.

Goggin, G. (2010). *Global Mobile Media.* New York: Routledge.

Haddon, L. (2006). Empirical studies using the domestication framework. In T. Berker, M. Hartmann, Y. Punie & K. J. Ward (Eds.), *Domestication of Media and Technology* (pp. 103-122). Maidenhead: Open University Press.

Harper, R., Regan, T., Rouncefield, M., Rubens, S., & Al Mosawi, K. (2007). *Trafficking: Design for the Viral Exchange of Digital Content on Mobile Phones.* Paper presented at the 9th international conference on Human computer interaction with mobile devices and services, Singapore.

Hjorth, L. (2008). Being Real in the Mobile Reel. *Convergence: The International Journal of Research into New Media Technologies, 14*(1), 91-104.

Höflich, J. R. & Hartmann, M. (Eds.) (2006). *Mobile Communication in Everyday Life: Ethnographic Views, Observations and Reflections.* Berlin: Frank & Timme.

Hsu, C.-L., Lu, H.-P., & Hsu, H.-H. (2006). Adoption of the mobile Internet: An empirical study of multimedia message service (MMS) *Omega, 35*(6), 715-726.

Hung, S.-Y., Ku, C.-Y., & Chan, C.-M. (2003). Critical factors of WAP services adoption: an empirical study. *Electronic Commerce Research an Applications, 2*(1), 42-60.

Kraut, R. E., & Fish, R. S. (1995). Prospects for video telephony. *Telecommunications Policy, 19*(9), 699-719.

Larson, R. & Csikszentmihalyi, M. (1983). The experience sampling method. In H. T. Reis (Ed.), *Naturalistic Approaches to Studying Social Interaction* (pp. 41-56). San Francisco: Jossey-Bass.

Latour, B. (2005). *Reassembling the Social: An Introduction to Actor-Network-Theory.* Oxford: Oxford University Press.

Lee, S., & Kwak, D. K. (2005). *TV in Your Cell Phone: The Introduction of Digital Multimedia Broadcasting (DMB) in Korea.* Paper presented at the Annual Telecommunications Policy Research Conference. Arlington, VA.

Lehtonen, T. (2003). The Domestication of New Technologies as a Set of Trials. *Journal of Consumer Culture, 3*(3), 363-385.

Leung, L. & Wei, R. (2000). More than just talk on the move: uses and gratifications of the cellular phone. *Journalism and Mass Media Quarterly, 77*(2), 308-320.

Ling, R. (2004). *The Mobile Connection: The Cell Phone's Impact on Society.* San Francisco, Oxford: Elsevier/ Morgan Kaufmann.

Marez, L. de, Vyncke, P., Berte, K., Schuurman, D., & De Moor, K. (2007). Adopter segments, adoption determinants and mobile marketing. *Journal of Targeting, Measurement and Analysis for Marketing, 16*(1), 78-95.

Noll, A. M. (1992). Anatomy of a failure: Picturephone revisited. *Telecommunications Policy, 16*(3), 307-316.

O'Hara, K., Mitchell, A. S., & Vorbau, A. (2003). *Consuming video on mobile devices.* Paper presented at the SIGCHI conference on Human factors in computing systems. San Jose, CA.

Palen, L., Salzmann, M., & Youngs, E. (2000). *Going Wireless: Behavior & Practise of New Mobile Phone Users.* Proceedings of the ACM conference on Computer supported cooperative work. New York, NY.

Parsons, H.M. (1974). What happened at Hawthorne? *Science*, 198, 922-932.

Pedersen, P. E. & Nysveen, H. (2003). *Usefulness and self-expressiveness: Extending TAM to explain the adoption of a mobile parking service.* Paper presented at the Bled Electronic Commerce Conference. Bled, Slovenia.

Peters, O. & Ben Allouch, S. (2005). Always connected: A longitudinal field study of mobile communication. *Telematics & Informatics, 22*, 239-256.

Quandt, T. & von Pape, T. (2010). Living in the mediatope. A multi-method study on the evolution of media technologies in the domestic environment. *The Information Society, 26*(5), 330-345.

Rogers, E. M. (2003). *Diffusion of Innovations.* New York: Free Press.

Ruggiero, T. E. (2000). Uses and Gratifications Theory in the 21st Century. *Mass communication & society, 3*(1), 3-37.

Schuurman, D., de Marez, L., Veevaete, P. & Evens, T. (2009). Content and context for mobile television: Integrating trial, expert and user findings. *Telematics and Informatics, 26*(3), 293-305.

Shin, D.-H. (2007). User acceptance of mobile Internet: Implication for convergence technologies. *Interacting with computers, 19*(4), 472-483.

Silverstone, R. & Haddon, L. (1996). Design and the domestication of information and communication technologies: Technical change and everyday life. In R. Silverstone & R. Mansell (Eds.), *Communication by Design: The Politics of Information and Communication Technologies* (pp. 44-74). Oxford: Oxford University Press.

Taylor, A. S. & Harper, R. (2003). The gift of the gab: a design oriented sociology of young people's use of mobiles. *Journal of Computer Supported Cooperative Work, 12*(3), 267-296.

Venkatesh, V., Morris, M. G., Davis, G. B., & Davis, F. (2003). User Acceptance of Information Technology: Toward a unified view. *MIS Quarterly, 27*(3), 425-478.

Wang, C.-C., Lo, S.-K., & Fang, W. (2008). Extending the technology acceptance model to mobile telecommunication innovation: The existence of network externalities. *Journal of Consumer Behavior, 7*(2), 101-110.

Wei, R. (2008). Motivations for using the mobile phone for mass communications and entertainment. *Telematics & Informatics, 25*(1), 36-46.

Williams, R. & Edge, D. (1996). The Social Shaping of Technology. *Research Policy, 25*, 856-899.

Wirth, W., von Pape, T., & Karnowski, V. (2008). An Integrative Model of Mobile Phone Appropriation. *Journal of Computer-Mediated Communication (JCMC), 13*(1), 593–617.

Wirth, W., Karnowski, V., & von Pape, T. (2007). How to measure appropriation? Towards an integrative model of mobile phone appropriation. In T. Hess (Ed.), *Ubiquität, Interaktivität, Konvergenz und die Medienbranche. Ergebnisse des interdisziplinären Forschungsprojekts intermedia* (pp. 83-105). Göttingen: Universitätsverlag Göttingen.

Wu, Y.-L., Tao, Y.-H., & Yang, P.-C. (2007). *Using UTAUT to explore the behavior of 3G mobile communication users.* Paper presented at the 2007 IEEE International Conference on Industrial Engineering and Engineering Management. Singapore.

The Appropriation of Mobile TV through Television Preferences and Communication Networks

Julien Figeac

1   Introduction

Six years after the marketing of Mobile television (TV), few French people frequently use multimedia cell phones to watch the fifty-channel broadcast. Some users associate this service with their daily media practices. This choice can be explained by two factors. First, when the television at home is unavailable, teenagers can watch television programs by using mobile phone in their bedrooms. This factor explains the use of this service in residences. Second, many users adopt this service because they use public transportation systems to commute to their workplaces. In addition to listening to music or reading, these users want to be entertained by watching television when travelling. Television viewing is favored over the other forms of media consumption, and the users want to try this new entertainment tool to view their preferred media during their daily commutes.

In this article, we are interested in this second case because users must negotiate two important constraints to watch Mobile TV when commuting. First, their television programming must be adapted to mobile phone screens. For example, can a TV series fan watch the latest episodes by using his telephone? Will the size of the screen, lifespan of the battery and telephone network connection enable him to follow an episode under good conditions? Will these technical constraints encourage him to use Mobile TV less often, choosing shorter programs, such as news shows? Second, the television preferences must be adapted to travel constraints. The user must evaluate whether the duration of an episode corresponds to their trip time. Also, trips between two subway stations and transfers in public transportation are problems for telephone network connections that are linked to underground transport.

Several studies have dealt with these questions. These studies have shown that numerous constraints prevent users from viewing their preferred television programs on mobile phones and while travelling. The small screen size of mobile phones (Chipchase, Yanqing, & Yung, 2007), the short time slots in which they are used and the noise of urban environments (Södergard, 2003) prompt users to select programs that are not their main television preference. These constraints, linked to Mobile TV, contribute to polarizing this televisual consumption towards the news because these programs can be watched over a short period of time (Knoche & Mc Carthy, 2005; Oksman, Noppari, Tammela, Mäkinen, & Ollikainen, 2007). This factor explains why television programs that are not based on a structure of intrigue are at the center of this televisual consumption (Figeac, 2007). These programs have neither a real beginning nor a real ending. They do not tell a story, and a user can watch the shows even if the beginning was missed. The viewer can stop watching the news whenever necessary without being frustrated in not knowing the end. Consequently, users tend to favor television content without a structure of intrigue because they can watch them for short moments during their trip. We conclude that the viewers choose their favorite television programs based on a narrative structure that is adapted to the utilization constraints related to daily commuting.

In this report, we describe how users show reflexivity in adopting Mobile TV and adapt their choice of television consumption to mobility constraints. We cannot be satisfied in saying that the viewers use this service to avoid being bored, from a desire to relax or out of simple curiosity or habit without any particular preference (O' Hara, Mitchell, & Vorbau, 2007). The users show reflexivity in selecting a TV program that can be watched on the small screens of their mobile phone over very short periods while commuting. The users denote a "focus, a suspension, a pause on what is happening" (Hennion, 2004) to select a specific television genre (Corner, 1991) and to evaluate the selected TV program. Through this reflexivity, they evaluate which program is practical to watch in the current circumstances. They evaluate which television preferences are adapted to this new screen and which ones correspond to the mobility constraints. Thus, they review and renew their television preferences based on these evaluations.

Contrary to a structuralist vision of media practices, we will show in this article how users assume control of the renewal of their media practices and how the practical circumstances and the ecological dimension of daily activities intervene in this dynamic of media consumption. We will show how technical mediations also play an active role in the daily renewal of media practices. Indeed, the principal constraint of use encountered when travelling is phone network availability. We will show how users negotiate this constraint as opportunists by transforming it into a resource that enables them to explore new content and new forms of entertainment.

1.1 The Structuring Effects of an Invisible Technical Mediation

In this article, we study a specific usage constraint/resource: the availability of a phone network. The connectivity of the network is constraining because it makes the use of Mobile TV possible. The unavailability of the phone network can force users to switch off a program, particularly in underground public transport. This constraint is specific because it refers to "worked urban environments" (Licoppe & Levallois-Barth, 2009) that access this mobile service usage.

To observe its effects on the appropriation of Mobile TV, we must refer to "ubiquitous computing" (Weiser, 1991) because this work preceded, from a theoretical point of view, the current developments of this service. According to Weiser, technologies must be invisible: they must be embedded in usage environments and must work in the background of the user's awareness. This invisibility of technical mediation can be observed in the use of Mobile TV because the phone network waves are not perceptible to users. The mobile phone manages the connection to this invisible technical network. In addition to this invisibilization of technologies, "ambient intelligence" (Dey, Salber, & Abowd, 2001) proposes to develop technologies that are able to determine by inference what a user does in a context or tries to do in order to offer the best-adapted service. Mobile services are not (yet) able to produce such opportunities of usage in proposing the relevant service to use in conjunction with the activity of the user, the type of phone network available, the force of the signal, the place, the time of day, etc. These adjustments must be made completely by the users. They must evaluate the media supports and the services available during their journey to determine which ones are relevant with regard to the practical circumstances.

This path of research can be explored in the prolongation of ubiquitous computing by observing how users resort to various opportunities of usage offered to them while commuting. The objective is to locate how the invisible work of technologies comes to augment urban environments (Harrison & Dourish, 1996; Dourish, 2006). The city can be divided into multiple territories, according to their technological equipment and the diversity of usage opportunities it produces. For example, railway stations form territories that are rich in opportunities because they are equipped with multiple communication networks: radio, telephone, GPS and, more recently, Wi-Fi hotspots. It thus becomes interesting to describe how users demarcate places (Harrison & Dourish, 1996; Dourish, 2006) in urban environments by territorializing their various media activities around the usage opportunities produced by technology. In this article, we will show how users demarcate places to use Mobile TV. Users who choose to watch Mobile TV rather than reading the press will evolve differently in their environments of mobility. The user who replaces reading the press with watching Mobile TV will move differently in urban environments. Indeed, he must evaluate the phone network reception in order to watch a televised program. To do so, the user will decode the graphic indicators of his mobile phone interface. Even if these indicators show the availability of Mobile TV at a precise moment, they do not provide any information on its availability during the trip. This technological resource does not enable the user to coordinate its use with the itinerary effectively, and the user must extract the indications that provide information on the (future) availability of Mobile TV from the urban environment. We will show that the indicators that the user will employ are the same as those that are used to follow the route (i.e., the name of the stations and place names). It is through the indicators that he will identify the places where it will be relevant to watch Mobile TV. Our observation point is thus the availability of the phone network to describe how users proceed to evaluate the connectivity of this technical mediation and, consequently, to evaluate the relevance of Mobile TV use in these practical circumstances. We will show how users demarcate places dedicated to the use of Mobile TV while describing how they index their television reception on the availability of this technical mediation. Through this double appropriation of Mobile TV and its territories of use, we will show how this service comes to supplement or to replace the media activities that the participants developed previously in their trajectory of utilization.

1.2 Methodology

The results we will present are extracted from a study in 2007 with a sample of 15 Mobile TV subscribers. This sample is primarily constituted of technophiles; i.e., 25 to 35-year-old college graduates with a professional occupation in a large French conglomeration (Paris). To observe their uses in situations of mobility, we conducted ethnographic observations by using camera glasses (Relieu, 2002). The users filmed their journeys on public transportation for one week. Our video database is based on the round trips between their residences and their workplaces. It represents approximately 80 hours of recording.

These video recordings enabled us to describe how travel directs Mobile TV uses and, reciprocally, how the methods of using this service direct the means of travel in urban spaces. To show engagement in this form of multi-activity, we first described the movements of users in public transport: their bust and head movements, their directed regards and their quick glances (Sudnow, 1972). While describing how a user stops looking in the direction of his telephone screen to look at the platform or the name of a station, we will show how his attention is diverted from the activity of following his route to focus on the activity of television reception. Then, while describing how he once again looks at the screen, which was kept within glancing range, we recompose the sequential organization of this form of multi-activity. We show how the distribution of visual engagements between the uses and the itinerary takes the form of a succession of logical operations, prefaced by body positioning.

To describe engagement in this form of multi-activity, we then showed the participants the video recordings of their activities. The objective was to have them explain their intentions and their tactics in using the method of self-confrontation interviews (Theureau, 2004).

This method was developed by ergonomists. It leads the user to discover the parameters, contained in the situational ecology, orienting his practices. Using the recordings, he can clarify his activities as he discovers them. The performativity of self-confrontation interviews constitutes a resource in understanding how media activities and television tastes are positioned. Interviews, questionnaires and logbooks are not very useful in understanding the ecological setting of media activities. The ethnographic observations are also too removed from the activities of the participants to make it possible to observe these settings. With these observations, it is not possible to understand how users articulate their reading of a newspaper and evaluating Mobile TV reception based on information from their mobile phone network indicators. Only the video recordings, coupled with self-confrontation interviews, show the action of these technological and ecological variables in the uses and in the reconfiguration of media activities.

2   The Reconfiguration of Media Activities as a Function of Communication Network Reception

The members of our sampling chose to watch Mobile TV while commuting, and they partially reconfigured their old media practices. We see this reconfiguration in examining the case of Linda, a 42-year-old receptionist. The ways in which she uses the media seem paradigmatic because the form of entertainment she chose was reading the newspaper, linked with watching television. To do this, she reorganized her media activities by alternating her reception of various media (press, radio and Mobile TV) according to phone and radio network receptivity. Indeed, during the 80 minutes between her residence and her workplace, she redirected her media activities four times, according to the reception or the unavailability of the telecommunication networks (see figure 8 at the end of this article). We now describe the sequential organization of her media practices. Ever since she has been able to watch television on her mobile phone, Linda does not perform the same ritual before crossing the threshold (Thibaut, 1994) of her residence because she no longer spends her travel time reading. She relies on her new telephone to replace reading magazines with watching Mobile TV.

Even if she no longer takes magazines with her, she still reads while commuting. She now picks up the free daily newspapers that other passengers have left on the seats of the subway. Consequently, in the morning, she enters the underground train searching for a seat where there is a free daily newspaper (see figure 1).

Figure 1:   Linda picks up "Matin Plus" before sitting down.

Figure 2:   Linda watches "Télématin".

Once she has found a newspaper, she does not read it immediately. She keeps it on her lap in order to read it later (Figure 2). In other words, she carries out a preparatory gesture (Datchary & Licoppe, 2007): she keeps this media resource and thus shows her intention to use it at another time (during this trip or later at her workplace or at home). It is important to determine when and under what circumstances the reading of the daily newspaper will occur.

Once seated, she immediately uses her telephone to watch "Télématin" on TV channel France 2 (Figure 2). During the self-confrontation interview, she says that she continues watching television, an activity that she had started during breakfast with her husband and children. Furthermore, she likes to immediately watch Mobile TV because her favorite television program (the TV news show "Télématin") begins at the exact time she takes the subway. In regularly taking this route, she has learned that she will not be able to watch Mobile TV beyond fifteen minutes when the train runs beneath Paris. Consequently, she watches Mobile TV as soon as she sits down in the train because she likes this TV news program, and she knows that it will not be possible for her to use this mobile service afterwards. Figure 8 (at the end of this article) shows how this receptionist watched Mobile TV for 13 minutes between the moment she took the subway (8:00 a.m.) and the moment she no longer had access to the phone network, just after "Télématin". We would like to understand how this user anticipates the unavailability of the phone network to switch off the Mobile TV before using another media. How is this anticipation deployed in regards to the practical circumstances of which it takes advantage?

We will answer this question using the video recordings. We will describe the sequential organization of this transition between these two media activities; i.e., the end of Mobile TV reception and the beginning of reading the daily newspaper. We will show which elements of this usage situation the receptionist perceives as relevant in operating this transition and how this transition is indexed to the availability of the phone network.

2.1 How the Problems of Reception and Media Activities Form a Pragmatic Test

To answer these questions, we transcribed the video recording of her trip to work (Figure 8). Once seated in the underground, she watches the televised news while keeping a free daily newspaper on her lap.

The connection to the program lasts approximately 2 minutes 40 seconds between the moment she launched it (01:50:07) and the moment the TV news appears on the screen (04:31:09). Then she watches this program for one minute before losing the connection (05:27:27). She needs an additional minute to restore it (06:34:13). She watches it again for 4 minutes and 44 seconds before losing the connection with the Mobile TV once again (11:18:17).

This second disconnection occurs 30 seconds after the underground leaves the Bibliothèque François Mitterrand Station (we will call it "BNF") (10:49:74). During the self-confrontation interview, she specifies: "Here, it cannot be picked up anymore (Mobile TV), so I listen to the radio". This statement proves that she had anticipated the unavailability of Mobile TV because she identified this area as a border area: "after the BNF, nothing can be picked up anymore." When she crosses this borderline on her way to the office, she says she stops using Mobile TV before launching the radio function of her mobile phone that she will listen to while reading a newspaper.

However, the transcription of this journey shows that she tries to connect once again to Mobile TV (11:35:06), even if the subway has left station BNF (10:49:74) for 45 seconds. In other words, she continued to attempt to access Mobile TV after she crossed this borderline. The launching of the connection lasts approximately one minute (12:17:10) until the moment she can watch the televised news again. She succeeds in watching it for seven seconds before the program is cut off (12:25:91), as she could have predicted. Only after this final attempt does she decides to listen to the radio with the FM tuner of her mobile phone (12:55:20). She activates the radio to listen to it while she reads her free newspaper. She opens the daily newspaper 20 seconds later (13:18:69) when the train arrives at Austerlitz Station (13:26:12) and the Mobile TV reception has become impossible. In these practical circumstances, she consequently reactivates the reading activity that she had left on stand-by 13 minutes before (00:25:08) in a preparatory gesture.

With the transcription of this sequence, we understand better why she was hesitant during the self-confrontation interview: "It picks up often, in fact. It picks up until the BNF. From there, after passing the BNF, it does not pick up near Austerlitz. Nothing can be picked up anymore." She knows that a connection with Mobile TV will be difficult at the BNF station, but she is not completely certain. However, she is certain she cannot launch Mobile TV when she is in the following station, Austerlitz. Why is it so difficult for her to identify the exact moment when the phone network is no longer available? Also, why does she continue to try every day to restore connection to this service in an area where the phone network reception is difficult?

We will answer this question by showing that she does not try to optimize these transitions between Mobile TV, the radio and the newspaper, even if she has identified the border where she will be forced to change media activities. The imprecision of this border is not a problem for her. On the contrary, this imprecision enables her to test the technological mediations that connect her to media activities: she continues 1) to handle her telephone to test it, 2) to understand whether the connection problems are related to her telephone or the phone networks, 3) to determine whether the telephone operator has deployed the Mobile TV network in this area, 4) to check if this service is usable in the underground systems in Paris and 5) to reevaluate the borders between Paris (where Mobile TV is unavailable in underground systems) and the suburb where she lives (and where this service can be consulted in overland transport). We now describe how she tests these technological mediations, which frame her media activities, before showing in the last part how this test leads her to requalify the locations of her media practices (de Certeau, 1980; Harrison & Dourish, 1996; Dourish, 2006).

2.2   At Time T of Network Indicators

Users can evaluate the intensity of the phone network with two indicators that are posted on mobile phone screens. The first indicates the type of phone networks to which the telephone is connected (e.g. EDGE, 3G, 3G+). The second indicates the signal intensity of this phone network.

Figure 3:   Loss of reception.

Using the video recordings, we can observe that Linda is connected to Mobile TV when these indicators show her that this service is available. When she is in an area where she knows from experience that Mobile TV reception is unavailable, she will try to launch this service if these indicators show her that the phone network is (temporarily) available. During the self-confrontation interview, she noted that she had identified that Mobile TV was unavailable between the BNF and the Austerlitz Stations. The phone network indicators, located at the top left of the screen (Figure 3), clearly indicated the loss of the signal to her when the TV news stopped (11:18:17). Just afterwards (11:32:45), these indicators showed her that the network was available again. Consequently, she tried to connect to Mobile TV (11:35:36) even if she knew that this service was usually unavailable at this location. This graphic information made connection to the service relevant under these circumstances.

When the service stopped after this last connection, the indicators showed that the phone networks were unavailable (12:25:91; Figure 4). The indicators contributed to dissuading her from using Mobile TV, and encouraged her to read a newspaper and use the radio function on her mobile phone.

Figure 4:   Loss of reception.

These indicators provide support to evaluate under which circumstances she can (or cannot) watch Mobile TV. However, they indicate to users that reception is not possible once the service is blocked, the program is stopped or the image is immobilized. This information appears too late to allow users to anticipate the unavailability of Mobile TV. The limit of these indicators is to reveal instantaneous information, relating the telephone signal reception to a precise moment without specifying the degree of receptivity of the signal in the upcoming minutes. This information would be useful for a user who would like to watch Mobile TV in the subway. An indicator could provide this information to the user if his trip was traced, the orientation of his journey predictable, his stops in the stations anticipated and the speed of his movements were calculated to indicate how long he will be in an area where Mobile TV is available. No technological mediation can fulfill this function (legally).

Consequently, users must seek other resources to anticipate the telephone signal reception to evaluate whether it is practical to start watching Mobile TV under particular circumstances. To anticipate the connectivity of this invisible technical mediation, environmental indicators that provide information on the telephone signal fluctuations are identified by the users.

2.3 In Light of an Invisible Technical Mediation

In addition to the indications given by network indicators, users seek indicators in the surrounding environment to help anticipate the availability of the phone network at moment T + 1. The users attribute the information given by the telephone interface to their location.

A large amount of information can be used to identify a location. Categorizing all of the architectural elements and other indicators that make it possible for a user to know their location during their daily commute is difficult. It is not possible for us to identify the various indicators by which Linda identifies station BNF. However, she can identify the places where she will be able to watch Mobile TV and to define the borders of where she must use another media by using the indicators. It is possible to isolate one of the parameters that enables her to identify the places of use for Mobile TV. The person who often takes the same route can move while watching Mobile TV. She manages her activities using the perceptible light around the screen of her telephone.

Like all users, Linda thinks that connection to Mobile TV is possible when public transport runs above ground. On the other hand, she thinks that the connection is interrupted when transport moves underground. Consequently, she relies on this indication, formed by light variations, in order to direct her media activities:

J.F.:   "And there you are watching TV..."
L.:   "Exactly, there the RER is outside, so I start watching TV."

If we take another look at her last connection to Mobile TV in light of this indicator, it appears that she begins this use (11:35:36) when the underground has been above ground for more than ten seconds (11:23:22). However, when the connection with the TV news program is established (12:17:10), the train has just entered a tunnel (12:16:22). Consequently, when the reception of this program stopped seven seconds later, this indicator dissuaded her from attempting the connection again because the train had moved underground. This example illustrates how external light is used as an indicator, which is both relevant and problematic, regarding phone network availability.

This example shows how variations in luminosity between the light of day and the semi-obscurity of tunnels can influence the deployment of various uses. When the user is focused on reception, he can easily locate these variations. In this example, Linda does not move her eyes from her telephone screen. She does not need to look outside to find her location because she can identify, through the rays that light the underground train, her departure from the BNF station. She sees in these light variations an indicator showing phone network availability. This supplements information given by network indicators, enabling her to anticipate Mobile TV connectivity at the precise moment T + 1; i.e., when the RER train heads above ground.

This example also shows how the information that users extract from external light is problematic. The fact that the RER moved above ground supplemented the information delivered by the network indicators, leading her into error because she could not watch the program for more than seven seconds. For this reason, the participants test the phone network availability every day because no indicator enables them to precisely anticipate these interruptions in connection. Uncertainty involved in the appropriation of Mobile TV shows us that the logic behind utilization proceeds from indications extracted from the situational ecology. However, these indications direct the course of action towards a model that differs from Gibson's theory of perception (Gibson, 1979). For Gibson, the information given by an affordance directs the single activity of a person who is moving. Consequently, the light rays of an affordance overlap a coupling of perception and action without the mediation of reflexivity being necessary; e.g., when a door handle is grasped.

In this case, variations in luminosity in the underground train direct two activities at the same time. The participants make use of these variations to exploit various uses and choose their itinerary. Luminosity transmits information on the phone network availability for the person who intends to manage these two activities simultaneously in an urban environment. This indicator is not useful for a user who is in a stopped train because the information from telephone network indicators is sufficient. The relevance of this indicator comes from this mode of involvement for this form of multi-activity. This indicator gives information to the participants who manage their journey on a peripheral level of attention around the Mobile TV screen on which they are focused. The light gives information to the user who awaits its appearance because he makes use of it as a reference mark that prefaces the orientation of its uses. Consequently, Mobile TV reception is embedded as a form of multi-activity because the indicators used to manage it are also used to manage the journey.

3   The Appropriation of Mobile TV Creates New Forms of Mobility

We described the commuting of a participant between her residence and her workplace. Because she has started watching Mobile TV, she does not wait for the underground train in the same station as that where she returns home in the evening. Before using this service, she waited for the underground train on the Boulainvilliers station platform. During this wait, she read a newspaper or a magazine. She let the trains that went in other directions of the underground line pass and took the one that stopped at her destination.

When she started using Mobile TV, she reconsidered her itinerary because the telephone network and the radio network were unavailable in this station. Now she enters the first train that stops at this station, no matter what underground line it serves. Then, she makes an intermediate stop in another station to take another train that stops at her destination. She selected this intermediate stop according to Mobile TV availability. Because the phone network is available in the above ground stations, she chose to change trains at the Champs de Mars station near the Eiffel Tower. This station offers a beautiful view of Paris, and she can take advantage of the panorama while watching her favorite television program (Figure 6).

Consequently, she thought of taking this initiative once she had the opportunity to watch Mobile TV. To use this service, she reconsidered her itinerary by modifying her habits. She reconfigured her trip around her attachment to television programs and phone network availability. This itinerary leads to increased mobility (Licoppe & Inada, 2006) through the use of new technologies. The journey is no longer calculated according to a purely strategic logic of connecting two destinations as quickly as possible. This search for performance is secondary compared to the pleasure felt watching television because she is entertained during the time spent in transport. This pleasure is why she now reorganizes her commuting time around the requirements that must be met in order to watch Mobile TV.

Figure 5:   Linda stands near the arrival and departure board.

Figure 6:   Linda is about to watch Laurent Ruquier's program.

Her engagement in this form of multi-activity made these two communication networks, the phone network and the transport system, converge at this particular place. The place where she stops to watch Mobile TV (Figures 5 and 6) is closely linked to this form of multi-activity. She stands near the arrival and departure board to be able to keep an eye on the arrival of the train (Figure 5). Through her preoccupation with the activity, which she will soon focus on (entering the train), she defines the temporal horizon of the activity of reception in this intermediate place. This example shows how the activities of reception and commuting overlap to form a multi-activity because the control of the first activity interferes with controlling the second and vice versa. As we have seen, the uses of Mobile TV modify the choice of the intermediate stations and the way in which these places are used. It is interesting to see how these places will retroactively affect the uses of this service:

> L. : "There I stop at the Champ-de-Mars station. In fact, I could have continued to the BNF station, but I preferred to stop. The weather is nice and all. And what's more, I could watch TV. [...] But it is not only because of that. It is also because the station is outside, and the weather is nice. So I can watch TV while soaking up the sun."
>
> J.F.: "And that is why you turned and faced the landscape?"
>
> L.: "Yes, it is. Here, facing the sun, facing the landscape, the Seine river... In spite of the noise of cars and the train...".

She appropriated this place (Harrison & Dourish, 1996; Dourish, 2006) because it allowed her "to watch TV facing the sun [...], facing the landscape, the Seine river... In spite of the noise of cars and the train...". Consequently, she did not choose this place only because the communication networks were accessible but also for its atmosphere (Thibaud, 1994). She said during the interview that the perception of the sunshine in this place appears to her, retrospectively, as the main reason she left the underground to continue watching TV on the platform: "the weather is nice and all, and what's more, I could watch TV." This example shows that the atmosphere of the places where Mobile TV is used is important in understanding its appropriation. Light is not only an indicator of phone network reception. The bright sunlight illuminates this place (Chelkoff & Thibaud, 1992). The sunlight makes it prominent on a practical level and makes it conspicuous on a subjective level as a place of exposure, allowing her "to get some sun".

It is necessary to show how technical mediations converge with the atmosphere of the places so that they can be suitable as places to use Mobile TV. If it is relevant to continue the paradigm of increased mobility while describing how the uses of information and communication technologies transform urban mobility, it is also necessary to describe how the emotional tonalities (Thibaud, 1994) of urban environments influence the uses of these technologies. These emotional tonalities do not have a direct impact on the televisual content choices. However, they enter into forming media tastes because they induce pleasures, which bind a user to a media practice and affect his value judgments: to appreciate Ruquier's television program "facing the sun, facing the landscape, the Seine... In spite of the noise of cars and the train... ".

4   Conclusion

Throughout these descriptions, we have shown how the uses of Mobile TV are linked to daily commuting. We have described how a user redefines her itinerary in public transport to watch Laurent Ruquier's program. As an opportunist, she exploits the availability of the phone network in stopping at the stations where Mobile TV reception is possible. She adapts to this constraint in preserving her old media practices to be used during the temporal interstices in which Mobile TV reception is not possible. This technical mediation partially structures her new itinerary by directing her towards certain stations, defining the phases and places (Harrison & Dourish, 1996; Dourish, 2006) where the uses of the various media become relevant. But these mediations fulfill this function if the user gives them this capacity, i.e., if he exploits them as an opportunist to renew his media activities during his journeys.

The video recordings show how this opportunist attachment with television preferences is carried out like a performance. The problems of Mobile TV reception do not form simple constraints. Even if it is considered to be prejudicial, it pushes the users to refer to their mobile phones and the environments where they wish to use them. It is through this referral process and the answers obtained that the appropriation of Mobile TV becomes a performance. As part of the mobile phone network, indicators do not enable them to effectively manage connections to Mobile TV. The users seek indicators that show the availability of the phone network in the urban environments. They exploit the name of the stations and the light variations to delimit and identify the places where Mobile TV reception is reliable. These ecological supports cannot be reduced to simple resources, exploited on a procedural level, in the prolongation of the situated action approach. The performance aspect is because the light of day is perceived differently through a new function. In the same way, the places, the distances between the stations and all this information that can make commuting unpleasant by their recurrence are exploited as invaluable indicators to direct the use of Mobile TV. Multi-activity (Datchary & Licoppe, 2007) is used to explain this embedding of the indications in the joint control of the uses and the activity of commuting. It is true that our method, based on video recordings, contributes to involving the participant in this state of concern. The user partly records what he sees with camera glasses. He is aware that the researcher can see from the recordings what he was looking at. Consequently, this method incites him to focus his attention on his telephone instead of looking around him and staring at the passengers. This method can amplify the performance engaged in this form of multi-activity. It can encourage users to quickly divide their attention between the media.

Through the concept of multi-activity, we have tried to highlight pragmatic methods of this performance. It seems to be a state of concern (Datchary & Licoppe, 2007) and a state of continual vigilance. In situations of mobility, the user never isolates himself from the surrounding world. He is vigilant, according to the two meanings of Goffman (1971): 1) to preserve his physical integrity in relation to others and 2) to preserve the normality of his physical appearance. To manage his engagement in this multi-activity, the user remains in a state of awareness to perceive the external signals that can lead him in directing his activity.

Even if the researcher's methodology has a real effect on the practices observed, we can still conclude this study with the following idea: this state of awareness, characteristic of the various forms of multi-activity, shows that the use of Mobile TV must be referenced to the peripheral activities occurring around its use. It is interesting to describe this performance, which consists of carrying out certain activities differently, particularly those that require little attention, such as media activities or cultural attachments. Through the study of Mobile TV uses, we can describe this performance of opportunistic attachment to television programs. This opportunism characterizes a form of attachment in which practical circumstances and technical mediations prompt users to select certain media preferences. This form of attachment is specific because there are not that many people using these services. Today, users have many different media preferences, which are constantly increasing with the growth of media. Consequently, to understand the adoption of new media and complementarities between various media, it is interesting to observe which media preferences they will select according to the utilization opportunities produced by new technologies and new services.

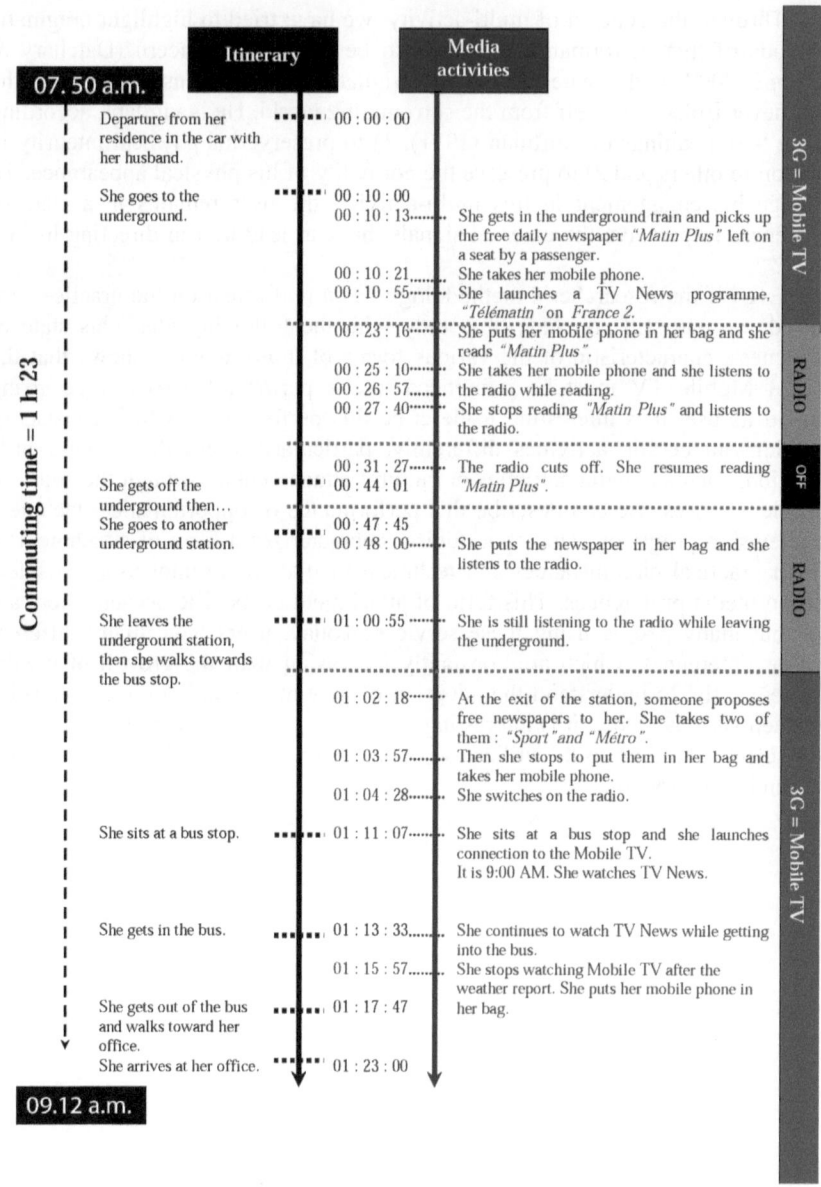

Figure 7: Linda's trip between her residence and her workplace

References

Certeau, M. de (1984). *The Practice of Everyday Life*. Berkeley, CA: University of California Press. (Original work published 1980)

Chelkoff, G. & Thibaud, J. P. (1992). L'espace public, modes sensibles: Le regard sur la ville. *Les Annales de la recherche urbaine*, 57 & 58, 7-16.

Chipchase, J., Yanqing, C., & Jung, Y. (2007). Personal TV: A Qualitative Study of Mobile TV users. *Proceedings of the 5th European Conference*, EuroITV 2007, Amsterdam, Netherlands.

Corner, J. (1991). Meaning, genre and context: the problematics of "public knowledge" in the audience studies. In J. Curran and M. Gurevitch (Eds.), *Mass media and society*. London: Edward Arnold.

Datchary, C. & Licoppe, C. (2007). La multi-activité et ses appuis: L'exemple de la "présence obstinée" des messages dans l'environnement de travail. *@ctivités*, 4(1), 4-29.

Dey, A. K., Salber, D., & Abowd, G. D. (2001). A conceptual framework and a toolkit for supporting the rapid prototyping of context-aware applications. *Human-Computer Interaction (HCI) Journal*, 16(2-4), 97-166.

Dourish, P. (2006). Re-Space-ing Place: "Place" and "Space" Ten Years On. *Proceedings of the 20th anniversary conference on Computer supported cooperative work CSCW'06*. Banff, Alberta, Canada.

Figeac, J. (2007). La configuration des pratiques d'information selon la logique des situations. *Réseaux*, 25(143), 17-44.

Gibson, J. J. (1979). *The Ecological Approach to Visual Perception*. Boston: Houghton-Mifflin.

Goffman, E. (1971). *Relations in Public: Micro-Studies of the Public Order*. New York: Basic Books.

Harisson, S. & Dourish, P. (1996). Re-Place-ing Space: The Roles of Space and Place in Collaborative System, *Proceedings of the 20th anniversary conference on Computer supported cooperative work CSCW'06*. Banff, Alberta, Canada.

Hennion, A. (2004). Pragmatics of taste. In M. Jacobs and N. Hanrahan (Ed.), *The Blackwell Companion to the Sociology of Culture* (pp. 131-144). Oxford, Malden: Blackwell.

Knoche, H. & McCarthy, J. D. (2005). Good News for Mobile TV. *Proceedings of Wireless World Research Forum 14*. San Diego, CA, USA.

Licoppe, C. & Inada, Y. (2006). Emergent Uses of a Location-aware Multiplayer Game: The Interactional Consequences of Mediated Encounters. *Mobilities*, 1(1), 39-61.

Licoppe, C. & Levallois-Barth, C. (2009). Configurer l'accessibilité des voyageurs équipés à des services mobiles multimédia: Le cas des publicités "augmentées" par Bluetooth dans le métro parisien. *Réseaux*, 4(156), 15-48.

O'Hara, K., Mitchell, A. S., & Vorbau, A. (2007). Consuming Video on Mobile Devices. *Proceedings of the SIGCHI conference on Human factors in computing systems*. San Jose, California, USA.

Oksman, V., Noppari, E., Tammela, A., Mäkinen, M, & Ollikainen, V. (2007). News in Mobiles: Comparing text, audio and video. *VTT Research Notes 2375*. Available: http://www.vtt.fi/inf/pdf/tiedotteet/2007/T2375.pdf [May 31, 2011].

Relieu, M. (2002). The "glasscam" as an observational tool for studying screen-based mobile phone uses and management of parallel activities. *Proceedings of the International Conference on Conversation Analysis* (ICCA-02). Copenhagen, Denmark.

Södergard, C. (2003). *Mobile television: Technology and user experiences*. Report on the Mobile-TV project, VTT Information Technology.

Sudnow, D. (1972). Temporal parameters of interpersonal Observation. In D. Sudnow (Ed.), *Studies in Social Interaction* (pp. 259-279). New York: The Free Press.

Theureau, J. (2004). *Le cours d'action: Analyse sémio-logique. Un essai d'une anthropologie cognitive située*. Toulouse: Octarès.

Thibaud, J.-P. (1994). Les mobilisations de l'auditeur-baladeur: une sociabilité publicative. *Réseaux, 12*(65),71-83.

Weiser, M. (1991). The Computer for the Twenty-First Century, *Scientific American, 265*(3.), 94-104.

Part III

Images and Representations
of Mobile Communication

Images and Representations of the Mobile Internet

Corinne Martin

1 Introduction

The success of the mobile phone has no equal in the recent history of ICTs (Information and Communication Technologies). According to the ARCEP,[1] the equipment rate of the French population grew up to 80% between 1994 and 2005. And it exceeds 90% in 2009. According to the sociology of uses and the way this sociology developed in France (Jouët, 2000), the mobile phone has been studied as a means of interpersonal communication and mediation with family, friends and work acquaintances. Its appropriation by a large majority of the French population has made it a banal object, perfectly integrated in the routines of everyday life's users (de Certeau, 1984). With the advent of images (MMS, Multimedia Messaging Services, in the beginning of the 2000s and then camera phones) and the mobile Internet, technological innovation has caused the mobile phone to evolve into a form of mass media. The first commercial start of the mobile Internet dates back to 2004, it was composed of optional subscriptions with pay-per-use (duration of the connection and/or quantity of data downloaed). At the end of 2007, the first "unlimited mobile Internet" packages arrived on the market, coincidentally with the first generation of iPhone in Europe. For the network operators, the goal was to foster the development of the mobile Internet, which at the time was lagging. In such a context of emerging uses (our study was carried out in spring 2009), we wanted to understand how representations of this innovation are constructed among these different social actors. Our project began as a reflection on resistance against the use of the mobile Internet in a polemical context concerning its new commercial offering. The representation of the mobile Internet produced is in a way an instantaneous, historically situated image of the reception of this innovation and it will ultimately evolve.

The goal of this edited volume is to analyze the advent of images and their impact on emerging uses of the mobile phone, including personal mobile televi-

---

[1] The French ARCEP (*Autorité de régulation des communications électroniques et des postes*) was founded in 1997. It carries out the regulation role on the telecommunications market and gives the official statistics on the French market. Access: http://www.arcep.fr/.

sion, photos, videos and the mobile Internet. To accomplish this goal in our chapter, we chose a slightly out-of-line approach; we shifted the focus. We chose to analyze the images created and produced through emerging mobile Internet uses. What are the social representations of the mobile Internet, its uses and its users? And how are they built? We limited this vast question to three main aspects.

In the first part, we will analyze the image of the mobile Internet user as it is constructed by various audience measurements. To do so, we relied on media reception studies that analyze the construction of television audiences (Méadel, 2009; Méadel & Bourdon, 2009; Méadel, 2010). With a critical analysis of the public discourse and statistics produced by various polling institutes, we will show how the image (still under construction) of the mobile Internet user is blurred and embellished in euphoric discourses. However, this image is converging in a single and consensual image. In the second part, we will study the image produced in the public space by the network operators and handset manufacturers in marketing new unlimited mobile Internet offerings. Analyzing controversies in the public sphere *via* the media, we will also show how aggressive these strategies are and how they rely on a certain level of symbolic violence. Similarly, the strategy of social distinction put into place by Apple and the network operators for the iPhone will be studied (an advertisement will be analyzed). In the third part, we will focus on the users; we will try to understand the social representations that they make of the mobile Internet through their limited use and their use of the mobile Internet in its early phase. By focusing on the early adopters of the mobile Internet and the first users of the iPhone (Ling & Sundsoy, 2009), our approach sheds new light on the issue of the uses. Through our sample of users (and even irregular users or non-users) of the mobile Internet, we discovered certain motivations and drivers as well as barriers. We lay the hypothesis that all of these motivations, drivers and barriers in the appropriation of a device allow us to explain jointly the use and non-use: use and non-use must not been analyzed separately (von Pape & Martin, 2010). Furthermore, mobile Internet use still appeared limited and the image of the computer for surfing the Internet remains dominant for the time being.

The methodology applied is combined. On the one hand, we analyzed the statistics of connections produced by various polling institutes as well as estimates made by experts. On the other hand, we carried out a study in two parts during the spring of 2009. The first part comprised a quantitative study using a questionnaire (N=262) for respondents between 18 and 23 years old: 111 students (1st and 2nd year in college) and 151 students of a professional high school. The questionnaire employed closed questions aimed at determining the use of the mobile Internet, including which device is used, which mobile Internet package is used, how often it is used, description of the uses, if it is not used and so on. Furthermore, it employed open questions aimed at constructing an inventory of the various obstacles for non-users of the mobile Internet and the possible motivations that can be drivers for the use of the mobile Internet in the future. The second part of the study relied on in-depth qualitative interviews with mobile Internet users for about an hour. The 15 interviewees were students in their final grade (*Terminale*) in a professional high school;[2] they were between 18 and 20 years old. According to their answers on the questionnaire, they were selected among the most important users of the mobile Internet in this sample: their use ranked from a several times a week to an everyday basis (two-third of the respondents never use the mobile Internet). The study was carried out during the spring of 2009; the interviews were focused on their experience with the mobile Internet to compare it with their use of the Internet on a computer.

2   The Mobile Internet User and Audience Ratings

2.1   A Representation Still Under Construction

The sociology of innovation shows how important the image of the user is throughout the innovation process. It is part of the social imaginary of technology that accompanies every innovation (Flichy, 2007a, 2007b; Scardigli, 1992). At the beginning of the innovation process, this social imaginary linked to the strategies of the main social actors: the state, the inventors and the producers, whose discourse is retransmitted by the media. We will be able to consider some traces of this discourse at the time of the early stages to the mobile Internet throughout its audience. This measurement reveals an image under construction, a blurred and not yet clear representation of the mobile Internet user; indeed, who is the mobile Internet user? The question may seem candid, but the different measurements that aim at describing this user do not provide the same image. These measurements by polling institutes, such as Ipsos, TNS Sofrès and Opinion Way, are completely different, as are the measurements by other

---

[2]   The school under study was a professional high school for vocational training preparing exclusively for a professional *baccalauréat* degree to have a profession immediately afterwards, here in the tertiary sector. (The degree is different from the general *baccalauréat* degree that allows the pursuit of studies at a university or in a business or engineer school. The *baccalauréat* degree is taken in the final grade in high school). The students of this school are in their majority coming from popular classes, some coming from the middle-class, and only a few of them from the upper classes. The choice of this school allowed us to limit the variable "high salaries" on the adoption of the mobile Internet. We would like to thank the Deputy Headmistress of this school for allowing us access to the school.

institutes that specialize in television audience measurements, such as Média-métrie, the French specialist in this field. To fully understand discrepancies in these measurements, a comparison with television audience measurement will be an instructive approach. In her analysis, Méadel (2009, 2010) reveals the essential stakes of using television audience measurements as a tool to build an audience. She shows how the "advertising equation" is defined and set up. This equation aims, through a series of complex operations, to set the commercial break price according to the television audience. There are three main operations: the reduction of data (from very detailed measurements of the viewer's behavior), the selection of significant variables (e.g., age, which enables us to define the target) and the translation of the spots into prices. All these operations result in a boundary object, the "standard-measurement", which will enable a media-planning strategy (the choice and evaluation of spots). However, the production of this common object of measurement needs an "agreement in competition" or, in other words, involves every social actor, despite the high competition, on one hand, between the broadcasters (i.e., television companies) and, on the other, the users (i.e., advertisers and their clients). This agreement relies on a necessary consensus, without which everything may become unmanageable. This is the reason why these different social actors created Médiamétrie in 1985, the standards institute of which they are shareholders. The institute's task is to produce this "standard-measurement". Méadel writes that this measurement is a major tool in advertisement because it offers indexes for the terms of the exchange: the TV viewer with his practices, his behavior, his environment and the space defined by a duration, a date, a position in the TV listings, a recurrence and so on (Méadel, 2009). However, she also shows how the system remains opaque and how the communication conditions of the measurement are heavily framed by Médiamétrie (Méadel, 2009). Consequently, this agreement can only be done with veracity; the data must be considered to be reasonably accurate, which leads to a concern for accuracy (fidelity to the real). On the same note, the most striking proof of the veracity of these measurements for the social actors resides unequivocally in the continuity of these measurements (Méadel, 2009). These measurements must be relatively solid. A last key element to keep in mind is that they are a representation of reality, a common standard.

2.2 A Blurred and Embellished Representation of the Audience

Our goal is to show how far we are from a common definition of this boundary object with the measurements of the mobile Internet use; no condition aimed at characterizing this "standard-measurement" has yet been fulfilled. We highlighted some significant variations between the different measurements we studied. These differences stem from a lack of consensus. To arrive at this result, we compared the results published during the same period (from January to November 2008) by five polling institutes: Médiamétrie for Mobile Marketing Association (survey carried out in March-April 2008); [3] TNS Sofrès (survey carried out in August 2008);[4] Ipsos (survey carried out from September to November 2008);[5] Opinion Way for IAB France (Interactive Advertising Bureau) (survey carried out in January 2008);[6] Ipsos Médias for AFMM (French Mobile Multimédia Association) (survey carried out in the 1st quarter of 2008)[7] (see table 1, p. 148).

It appears that a rigorous comparison of these statistical data is bound to fail. This is why the image of the mobile Internet user that has been constructed from this data is a blurred and changing one. First, the time of reference needed to be considered a mobile Internet user is not the same. A user is a person who connected to the mobile Internet once (at least), but the connection may have been made in the last 3 or 6 months, or even in the last month (Médiamétrie). Further, the population of reference is never the same (the French population, Internet users and mobile device owners). These variations seem significant. Let us compare TNS Sofrès' and Médiamétrie's measurements.

---

[3]   "11,3 millions de mobinautes en France", 07/16/2008. Available: http://www05.r7g.com/50071/e/new_2008_07_16_%20CDP%20panel%20mobinautes%20DC OM%20LOM.pdf [May 15, 2011].

[4]   "Avenir prometteur pour le M-commerce et en devenir pour le M-marketing", 11/17/2008. Available: http://www.tns-sofres.com/espace-presse/news/E00CE85007DA4BD185AF382D21 E5DD8C.aspx [May 15, 2011].

[5]   "Profiling 2008-V2", 01/12/2009. Available: http://www.offremedia.com/DocTelech/ Newsletter/DPIpsos12012009.pdf [May 15, 2011].

[6]   "Internet et la téléphonie mobile". Available: http://www.opinion-way.com/pdf/opinionway-iab-internet_&_la_telephonie_mobile-pres.pdf [May 15, 2011].

[7]   "Internet mobile en 2008: usages et comportements", 04/11/2008. Available: http:// www.afmm.fr/img/CP%20pdf/2009/CP%20Ipsos%20AFMM%2011avr08.pdf [May 15, 2011].

Table 1: Comparison of Audience Ratings Between Polling Institutes

Polling Institute	Period of the poll	Audience ratings	Population of reference	Duration of reference
Média-métrie/AMM	March – April 2008	11.3 million	Of the French population over 11 years old	Last 30 days
TNS Sofrès	August 2008	20% [it represents roughly 7.8 million]	Of the French population between 16 and 60 years old	Last 6 months
Ipsos	September – November 2008	3.3 million [it represents roughly 10.3% of Internet users]	Of a subgroup of 32.1 million Internet users	Last 30 days
Opinion Way/IAB	January 2008	33%	Of all Internet users	Not provided
Ipsos medias/AFMM	1st quarter 2008	25%	Of mobile device owners aged 15-50	Last 6 months

By extrapolating from the population data of the INSEE (*Institut national de la statistique et des études économiques*, official French Institute for Statistics), we find that there are 7.8 million mobile Internet users over the last six months according to TNS Sofrès, whereas Médiamétrie finds 11.3 million over the last month. Which is the fidelity to the real (Méadel, 2009) with such variations, which is the most accurate measurement? The contradictions are clear and we can only formulate hypotheses to try to explain these differences. For now, we are only able to see the tip of the iceberg. The details of these audience ratings and the complete results are still a jealously kept secret, as noticed by Méadel regarding the television audience (2009, 2010), whereas a consensus between all the involved social actors on the measurement has been found. In the case of the mobile Internet audience, the lack of communication of the detailed results and the lack of a methodology for measurement can be easily explained by the fact that there is no consensus on the measurement. We could say that the impossibility of comparison is a good thing for these polling institutes. Why? First, we are still facing an uncertain phenomenon because the mobile Internet remains at the early stages and users are not yet accustomed to it. The second reason, which is related to the first one, is that we have to face the fact that the number of users did not grow significantly until the beginning of 2009: the yearly study of the CRÉDOC (Bigot & Croutte, 2009) shows for the first time that the mobile Internet is "finally taking off", the 10% users' mark have been reached, although the figure have remained the same since 2004. Actually, the number of users did not grow fast enough for the network operators and figures representing these data in 2008 are not always welcome.

Moreover, we even found a discrepancy between two measurements by the same polling institute, Médiamétrie. There are 11.3 million mobile Internet users in March/April 2008 (see note 3) and 11.4 million at the end of 2009, whereas the mobile connections represented 0.2% of all Internet connections in June/July 2008 and had soared to 2% by the end of 2009.[8] How is this possible? What is taken into account for the measurement? The black box of the audience measurement revealed by Méadel and Bourdon (2009) for television seems the same for the mobile Internet.

Minges (2005) also noticed the same phenomena in Japan, where there is a lack of official data, especially for mobile Internet subscribers: "One reason is that some operators do not distinguish between mobile voice calls and calls to the Internet. In other cases, the number is probably considered too low and thus potentially embarrassing" (Minges 2005, p. 117). Indeed, the stakes are strategic. Why?

The reasons lie in the advertising equation studied by Méadel (2009, 2010). The ultimate goal is to determine the price of the advertisement that will come on the screen of the mobile phone. For the time being, the priority is to attract advertising companies so that they are willing to invest in this new media. The mission that inspires the AFMM, the French mobile multimedia association, is "to promote the Gallery offer to companies, administrations and authorities; to develop the notoriety and the usage of the Gallery services by the general public",[9] with Gallery being the mobile Internet portal provided by the three

---

[8]   Médiamétrie, "Année internet 2009", 03/10/2010; no more available on Médiamétrie's Website.

[9]   AFMM Website: http://www.afmm.fr/ All these quotations were available on the Website at July 15, 2009. No more available.

mobile network operators. The mission is clear: take part in the development of mobile Internet usage and make users and brands meet each other. The founding members of the AFMM are the three main mobile network operators (they share 95% of the French mobile market in 2008, according to the ARCEP, see note 1) and two associations promoting the digital economy and, more specifically, the digital edition, of content (Acsel, an association of the digital economy and Geste, a group of online services editors). Other members, such as the AMM, the French mobile marketing association and some MVNOs (mobile virtual network operators), have followed suit. Aiming to develop the mobile Internet, the AFMM logically ordered one of the audience ratings we analyzed. It appears that the aim is to show potential advertisers an embellished image of the audience. For instance, the rate of mobile Internet users (25%) is, methodologically speaking, biased and incresased because of the choice of the population of reference. Only mobile owners aged 15 to 50 are taken into account. By includ-ing younger users (with less money) and older (less interested) users, this rate could be lower. However, such variations are bound to disappear because we are reaching a unique measurement of the audience for a consensual image.

2.3   Toward a Consensual Image of the Audience?

The same AFMM is at the origin of the launch of a call for tender to select a supplier that will take responsibility for producing the unique measurement based on the same model of the "agreement in competition" that Méadel (2009) studied for the television audience. This measurement will concern all the social actors of the market: advertising clients, media and advertising agencies, adverti-sing production departments and mobile websites' publishers. Médiamétrie won this market (against the German Gfk) in June 2009. A first remark could be read on the AFMM website: "this measurement will strongly help the development of the mobile communication channel as a full media." This is clearly a sort of official recognition, which establishes and participates in the construction of another image of the media. It is not only an impersonal means of communica-tion, but can be, just like the Internet, put into the specific category of individual mass media. The AFMM asserts that this medium aims to be a "communication channel" for the advertising message.

Second, it is important to know that exhaustiveness is a characteristic of this measurement; it is developed both by census and by a panel (according to the recommendations of the GSMA, for a 'coherent' audience measurement throughout Europe; the Global System for Mobile Communications Association gathers all the actors of the communication at an international level). Development by census means that the whole population of mobile Internet users is the basis of the measurement and the data collected by the network operators (after having been made anonymous) will then be analyzed. These data include information on factors such as which website is visited as well as when, where and how often sites are visited. It gives a glimpse of an incommensurable quantity of information to manage by the polling institute Médiamétrie. The first task, which is colossal, is to reduce the data (Méadel, 2009; Méadel & Bourdon, 2009). In any case, we can conclude that this sole audience measurement, the result of an "agreement in competition", will give a solution to the advertising equation by defining the prices of the advertising market on mobile phones. The actors expect great things; the mobile marketing and m-commerce make all the brands (related to mass consumption) dream of advertising on the tiny screen of the mobile phone. The network operators are ready; they have already developed advertising production departments, without any real achievements thus far. Uses are not developed and the brands are very cautious about investing in this new media. There are also some deontological questions that have been raised.[10] Having analyzed the very first steps of the construction of the mobile Internet user, as defined by Internet audience measurement apparatuses, we will now concentrate on another image, the one produced and transmitted by network operators and handset manufacturers through their marketing strategies for commercial offerings on the mobile Internet.

3 The Image Constructed by the Marketing Strategies of Network Operators/ Handset Manufacturers

3.1 Aggressive Strategies and Symbolic Violence?

First, we will see how symbolically violent these strategies are. The following section will be dedicated to the analysis of the distinction used by Apple for its iPhone. First of all, we have to keep in mind that on the economic level, mobile communication is a highly strategic market with enormous stakes. In 2008, a business volume of 18.6 billion Euros was reached for all the market of the fixed

---

[10] See Licoppe and Levallois-Barth (2009), who write about the mobile marketing project in the Parisian Metro, where multimedia content could be sent to anonymous passers-by through a Bluetooth device. Reaching far beyond SMS marketing, we can easily imagine the interest that advertisers have in mobile marketing, but this particular project is currently suspended by a deontological problem: personal data protection, defended by the CNIL (*Commission nationale de l'informatique et des libertés*, the National Counsel for Informatics), and the conformity to the LEN, the French law on digital economy of 2004 (as an application of an European Directive).

phone, the mobile phone and the internet, according to the ARCEP (see note 1). This market is thus qualified as oligopolistic by the French consumers' association UFC-QUE CHOISIR, as three network operators share 95% of this huge market (UFC-QUE CHOISIR is one of the most influential among the French consumers' associations). We will show how an image of domination created by the aggressive marketing strategies of the network operators and the handset manufacturers can be seen. The starting point of the analysis is the amount of investments in advertising made by the network operators in order to sell their offerings (the offerings are not limited to the mobile Internet). Orange and SFR, the two main network operators, are among the first 10 companies within the ads market in France for 2007 and 2008, including all media (press, TV, radio, Internet and so on). [11] In 2008, Telecommunications sector was ranked 3rd just behind the leaders in delivery/mass distribution and automobile industries; this shows an important presence of advertisement for mobile phones on the entire French media panel. The means for commercial communication built by the three mobile communication providers have such an impact that it seems hard for the average citizen to escape such messages, except by boycotting the mass media. The second observation aims specifically at the unlimited offerings of mobile Internet services as they were launched in 2007 by the operating companies. Previously, the users had paid on demand (by the time they spent online or by the downloaded data). From a marketing point of view, these commercial offerings aim to foster the use of the mobile Internet.

Therefore, a flat rate for mobile Internet access was released, accompanied by controversy. Users quickly realized that this unlimited offer was actually restricted (limit of downloads). Most interestingly, SFR justified itself with the argument that the restrictions were a question of network sharing: "The 3G-network being mutualized between all users, SFR keeps the possibility of restricting the access of its users downloading more than 500MB of data monthly, in order to give access with optimal conditions to other users. The downloading of more than 500MB is not considered a reasonable use, as it is damaging the quality of the network for other users". (Excerpt of the general sales conditions of SFR, quoted in *L'ordinateur individuel*).[12] Why is it possible to speak of symbolic violence? For Pierre Bourdieu, symbolic violence relies on the imposition of perception categories on the social world. The dominated class takes part in domination as soon as there is a lack of knowledge. This violence applies itself precisely when it is ignored as violence (Bonnewitz, 2007). It appears

---

[11]   Media poche from Havas Media France. Available: http://www.media-poche.com/ [February 16, 2010].

[12]   "Internet mobile, c'est 'enfin' parti!", *L'ordinateur individuel* (02/01/08). The French daily press is available on the international database *Factiva*.

clearly that the network operator tries to impose its perception categories. Whereas the Internet is supposed to be unlimited, it is the user who is accused and sanctioned by the restriction of use. The accusation is explicit with the "use considered as unreasonable". Then, after recognizing the user as guilty because he is "damaging the quality of the network", the discourse shifts to an appeal to moral values. Users paying for an individual subscription are supposed to show solidarity by mutualizing the network. The image created by the dominant shows the intensity of the gap between the two present social actors: on the one hand, network operators that economically capture their subscribers[13] and on the other hand, users who must behave and respect moral values. In the case of economic versus moral values, the adopted scheme verges on caricature. It is also important to see that the two other network operators mimicked the same practices, but that all of them will soon be compelled to unbridle their Internet offerings by state intervention, following the counter-powers used by some consumers' associations. The AFUTT (The French association of telecommunication users) noted that the amount of user complaints related to mobile bills increased by more than 32% between 2007 and 2008.[14] In its study, the INC (*Institut national de la consommation*, the National Institute for Consumption is a public organization) also noticed bills going up to hundreds or thousands of Euros.[15] Finally, the French Secretary of State for Industry had to intervene by instructing network operators to set up alerts to be sent to customers after they reach a certain threshold of use.[16] It appears now that there is a need for the State to regulate and limit the power the network operators have over the consumers. Perfecting this image of almighty network operators restraining their users, some of them have recently prohibited the use of VoIP (Voice over Internet Protocol) software throughout Europe – the free use of voice on the Internet – on their 3G networks. How can these marketing strategies be considered anything other than war-like steps using, by their very strength, symbolic violence? The image of domination or almightiness is evident, facing a user who is reduced to a passive consumer, who cannot be considered a social actor or citizen, except when his moral sense is called for in mutualizing the network. The power struggle is obvious. But the controversy at the launch of the mobile Internet also reveals a paradox: the network operators limited uses whereas they wanted to foster them. We will now

---

[13]   "Vendre par forfaits, ou ceux qui capturent leurs abonnés", *Libération* (04/17/09). Available on Factiva.
[14]   "Les factures des mobiles passent mal", *La Tribune* (03/26/09). Available on Factiva.
[15]   "Multimédia mobile: gare aux factures!", *Le Figaro* (02/24/09). Available on Factiva.
[16]   "Les plaintes se multiplient contre le montant des factures liées à l'internet mobile", *Le Monde* (04/05/09). Available on Factiva.

analyze the iPhone as a case study to show how another image can be built, the image of social distinction.

3.2   The iPhone: Brand and Distinction Strategy

We will demonstrate how Apple's strategy is based on distinction, which pervades their economic and technological strategy (business model, technological systems, industrial design and so on). West and Mace (2010) explain the success of the iPhone by analyzing Apple's strategy, which was very different from other players in the telecommunications market. For these authors, there were two keys to this success. First, "rather than trying to recreate the Internet, Apple focused on re-creating the mobile phone to make it a good client to the already-mature ecosystems of the wired web" (West & Mace, 2010, p. 1-2). (Indeed, the efforts to create a second Internet for mobiles were doomed to fail, as the failure of the WAP, the Wireless Application Protocol, has shown). Second, Apple leveraged its system capability – especially its iTunes content ecosystem and other elements of its system's integration competencies – to establish a permanent position of value capture in the mobile phone industry (West & Mace, 2010). Furthermore, Apple always tried to distinguish itself from other competitors, even when it was a computer manufacturer. Finally, two elements are fundamental: ease of use and industrial design, both of which Apple used as sources of advantage (West & Mace, 2010).

As Kapferer (2008) pointed out, the brand is not the product; it is the meaning and it defines identity in time and space. The brand provides different functions, such as personalization and it is important for the individual to see that he is comforted in his self-concept, which Kapferer (2008) described as the image that someone gives to other people. As such, the people owning Apple products develop a real rhetoric underlying their adherence to the brand. We are in a value system, which almost amounts to being part of a community. That is for the distinction image that the brand always seeks to have *via* its owners. The distinction strategy also relies on two points. First, Apple launched a new business model with an exclusive agreement with network operators; Apple brings them new customers and new revenues in exchange for its control of downloaded content (West & Mace, 2010). In France, the network operator was Orange. However, this exclusive agreement has been revoked by the National Competition Council (*Conseil National de la Concurrence)* in December 2008; the court ruling was confirmed by the French Court of Appeal in February 2009 and the iPhone can now also be sold by the other network operators. Second, the iPhone is a closed system, which allows it to both create and capture value. By encouraging the supply of third-party applications, Apple bypasses the network operators (West & Mace, 2010); we can say it makes a real achievement by operating disintermediation with its interference *de facto* or by force in the loop network operator/customer relation.

Consequently, the App Store represents this direct link with the user[17] and its success is interesting to watch. Apple says there are more than 100,000 applications (as of the end of 2009) and it is easily understandable that this has become the core of the iPhone's advertising rhetoric. Let us analyze the message that appeared in one of the advertisements:[18] *"If you need to find a cab in an unknown town, there is an app for that... or to look at your budget this month, there's an app for that... or to repair a rickety shelf, there's even an app for that... In fact, there's an app for just about anything... only on the iPhone."* Music is played with a voice-over and the image focuses on the finger playing with the touch-screen and the apps – the colored buttons that characterize the iPhone. Calling a cab seems to be a self-evident use of a mobile phone; managing their budget comes from a computer, but controlling the uprightness of a shelf comes as a surprise. One can think about a simple gadget, with the screen becoming a waterlevel to be put on said shelf. However, it also refers to the idea of the Swiss Army knife that emerged as a social imaginary during the beginning of the mobile Internet in 2004 (Martin, 2007). This universalism makes the object powerful and almost magical: *"there's an app for about anything".*[19] Apple's rhetoric also refers to the real Internet that Steve Jobs promised (in opposition to the portal sites of the network operators). The iPhone is not only smart, but it can do anything and it is easy to understand how the desire of possession can be satisfied by the personalization it gives (Kapferer, 2008). To reinforce his self-concept, the individual will not stop desiring to possess – through appropriation, we might say – the qualities of this object in order to become almighty in the management of his everyday life. Finally, *"only on the iPhone"* attests that only the iPhone is capable of doing such things. Here, the distinction strategy is fulfilled; it is this object alone that one must possess. At the same time unique and universal, it arrives as a supreme privilege bound to distinction. Pierre Bourdieu (1984) showed how the consumption of cultural goods is enshrined in an aspiration for social distinction and results from the dominant/dominated struggle and the will of these dominant groups who accu-

---

[17] It is important to note that P2P is impossible on the iPhone, as you have to go through iTunes, the paying platform of Apple. That is the reason why some interviewees in our survey (*cf.* 3rd section) explained that they still used their computers to download free music or had to use cracked software.

[18] Broadcast on French television during the year 2009. Available: http://www.youtube.com/watch?v=XoPZnz9vxwk [May 15, 2011].

[19] An article of *Libération* "L'iPhone, joindre l'inutile à l'agréable" (11/23/09) denounces the "stupidity of the phone apps" while acknowledging that it "doesn't harm their success".

mulate symbolic capital. From the most legitimate and distinguished practices to the illegitimate and the vulgar ones, the consumption of cultural goods is a classifying consumption. It seems that the iPhone is currently appealing to this sense of legitimacy. This seems to be the impact that Apple's marketing strategy has desired from the start. But how long will this last? If the iPhone is to become a mass product (the iPhone represents 17% of the smartphones for the Idate, quoted by JRC, Joint Research Centre of the European Commission and the Institute for Prospective Technological Studies, 2010), it will lose its distinction power and simply become ordinary, like the mobile phone. Will we find the iPhone in the popular classes? Will we find it in emerging countries?[20]

This hegemonic image suggested by the advertisement quoted above can also be applied on the manufacturers' side, as they seem to be reduced to imitating Apple's new products. As the image of perfection that must be imitated, the iPhone was voted *Invention of the Year* in 2007 by *Time Magazine*.[21] All of these elements participate in the creation of the euphoric discourse that produces a better image and leads to what we can call "iPhone-mania". This discourse is part of the social imaginary accompanying the innovations (Flichy, 2007a, 2007b) and aiming to build and confirm this success story.

If we now focus on the uses, the iPhone is recognized as having changed the conception of the mobile Internet. West and Mace (2010) describe the Apple's browser as the "killer app" because it provides a browsing experience closer to that provided by personal computer than any previous mobile phone. Ling and Sundsoy (2009) showed how the iPhone was the device that most encouraged surfing on the Internet, compared to other devices in Norway in 2008 (even if it is not the only one). iPhone users use the Internet more than other users do and, furthermore, they have different patterns of use. The authors put forward several explanations, among which the socio-demographics of the iPhone users, the nature of the device and the nature of the subscriptions. However, they also evoke a kind of self-fulfilling prophecy; within the context of the marketing of such a "mobile Web terminal", the iPhone users have to justify their purchase. Indeed, it works like a virtuous circle. To conclude with an iPhone case study, the iPhone represents two thirds of the mobile connections of a panel of websites in France in 2009 (Médiamétrie, see note 8). However, we have to keep in mind that mobile connections represent only 2% of all the Internet connections in 2009 (Médiamétrie, see note 8).

---

[20]   For a qualitative study on mobile Internet use (but not the iPhone) in low-income communities in urban South Africa, see J. Donner & S. Gitau (2009).
[21]   Available: http://www.time.com/time/specials/2007/article/0,28804,1677329_1678542,00.html [May 15, 2009].

After studying the image transmitted by the network operators and the handset manufacturers to the audience through their marketing strategies – aggressive or distinction strategy – we will now turn to the image the user has of the mobile Internet. Whether he is using it or not, what social representation does he construct of this technological innovation?

4   Users' Social Representations of the Mobile Internet

The goal of this final section is to understand which mental and social images are given to the different uses of the mobile Internet. The French-speaking sociology of uses (Perriault, 1989; Breton & Proulx, 2002; Jouët, 2000; Mallein & Toussaint, 1994) showed how the appropriation of an artifact requires not only the instantiation of a physical, corporal relationship with the technical object, but also a mental relationship allowing the representation of the object as such. Through this appropriation process, the user constructs the standards, social meanings and legitimations of his practice. To understand how these standards are built socially and mentally as an integral part of the image of the mobile Internet for users, we will use the concept of social representation as defined by social psychology (Moscovici, 2001; Jodelet, 1991; Moliner & Tafani, 1997). A social representation presents itself as an ensemble of cognitive elements (opinions, information and creeds) related to a social object. A social representation is a socio-cognitive construction; it is depicted both as the process and as the result of social construction. It has a sociological texture and is a subject's individual production submitted to the rules of the cognitive processes. It is a social representation because it is collectively generated through social interactions and consequently it is shared by the members of a social group. This representation is socially useful because it takes part in the construction of reality. We will then try to discover users' social representations of the mobile Internet through a qualitative study in the form of in-depth semi-directive interviews. We will also use some results of our questionnaire to complete the image of mobile Internet by non-users and to discover some barriers to the use (*cf. supra*). First, the image of the use of the mobile Internet is an image of limited use; second, the image of the computer associated to the Internet is still predominant.

4.1   A Limited Use

As a result of the convergence boosted by the providers, the image of a unique object for all use should have appeared, but, in fact, many existing objects are supposed to do everything. The user then finds himself paradoxically in front of an excessive commercial offering. Additionally, his answer to this variety of available devices is a diversity of uses. The extreme personalization of these new uses is remarkable. Having the choice between various competing devices and applications, users develop a complex logic of how to choose between and combine them in their everyday life. This logic of uses (Perriault, 1989; Mallein & Toussaint, 1994) can be described as a rationality of uses (Caradec, 2001). As we will show, this rationality is based on the personal analysis of many criteria: technical, ergonomic aspects, material, economical, affective constraints, or constraints linked to the personal history of the user. This arbitration is neither absolute nor definitive; on the contrary, it is a relative one because it is continuously realized in that it evolves in time with the different devices that are available, which leads to a reorganization of the criteria.

Lena (19 years old, her father is a foreman, not a smartphone, only MSN option) says she often takes photos and videos and sends them to her friends: *"The function of a mobile phone changes with the other device you have."* Let us take another example. Thomas (19 years old, his father has a restaurant, 12th grade, professional high school, smartphone Nokia, no mobile Internet plan but Wi-Fi connection) is a fan of video games and therefore has chosen his latest mobile phone according to this aspect: his choice was a Nokia designed for video games, specifically adapted for this new use. It is clear that the "reasons" are adding up and are affective and economical. Thomas can engage in his passion and play at any time (during a break, while sitting in public transportation, in his bed, etc.) and he avoids buying the latest video game console in order to have the state of the art material (he'd rather buy a new mobile phone, which he would have done anyway). He can then download free video games through the Wi-Fi connection of his mobile phone. His use of the mobile Internet is limited almost solely to downloading video games (he does not even have an Internet subscription; he connects only at home through Wi-Fi connection). This limited use is the second most striking point of these emergent uses of the mobile Internet; indeed, usage is often limited. That is what the respondents using the mobile Internet expressed in the survey. Let us listen to Benjamin (20 years old, his father is an ex-cameraman retired, he lives with his girlfriend, 12th grade, professional high school, iPhone with flat rate mobile Internet): *"When I am in a remote location and there is no computer, I can still surf on the Internet."* His mobile phone simply and temporarily compensates the lack of a computer, with the idea of being stranded in a remote place. A similar case is Katia (19 years old, her father is a blue-collar worker, her mother died, she lives alone with her sister, 12th grade, professional high school, iPhone, flat rate mobile Internet for iPhone). Katia is very proud of her iPhone because she likes being the *"first one"* to have something: *"My mobile phone is my computer"*, she says, but she adds later on *"I know I only have this, so I content myself with it."* The expression *"contenting oneself with"* is revealing; Katia indeed does not own a computer, as she is living alone with her sister. The image of the mobile Internet is built in a sort of hierarchical organization with the computer, the primary use of which is always Internet access. Thomas (19 years old, his father has a restaurant, 12th grade, professional high school, smartphone Nokia, no mobile Internet plan but Wi-Fi connection) explains: *"So let's say I'm at a gas station, I'm on holidays, there is a Wi-Fi access, I turn the Wi-Fi on, I turn MSN on, ok, but when I'm home, the computer is on, that's it."* The computer is on at home. It then appears that the use of the mobile Internet is 'limited', being used to offset the lack of the computer; likewise, it is used to have real-time access to information (news, sports) at school and it is also used to kill time, as the playful dimension is important. Then, perpetual contact (Katz & Aakhus, 2002), the same use as in voice telephony, reappears when it is important to keep in touch (Ling & Yttri, 2002) with one's network (Facebook, MSN, etc.), but it is still an irregular use, an emerging complementary use of the use of the computer and it is far from being able to be qualified as a "mobile-centric internet use" (Donner & Gitau, 2009).

We can complete this image with the respondents of our questionnaire who do not use the mobile Internet at all. Their discourse reveals an image that can be qualified as stereotypical because it is slightly disconnected from the reality of practice (they did not try or test the mobile Internet for a long enough time). This image is built with several dimensions linked to different categories of *"reasons"*: it is too expensive; *"it will use up all my credit"*. This is the typical answer of young people who do not have flat rate Internet. When their mobile phones are not smarphones with a touch screen, its ergonomics seems to them to be completely unsuitable for these uses. A European study that focused on mobile search noticed that cost and usability issues are among the main barriers for adoption of these services (JRC, 2010). Furthermore, the image of their phone is still linked to the interpersonal communication for the respondents to our questionnaire: *"I'm using my phone to give a phone call or send text messages"* is a typical answer. Moreover, they do not have any desire to use the mobile Internet because the absence of *"need"* is obvious for them: *"In any case, I got a computer at home!"* This discourse became a leitmotif throughout our questionnaire in 2009. We will now try to understand what it means through the in-depth interviews of our qualitative survey.

4.2   The Image of the Computer

Put simply, habits are difficult to overcome, the weight of habits is a hard one and the computer is still the primary access point to the Internet. Accessing the Internet through a computer represented close to 98% of all connections in 2009 and accessing it through a mobile phone represented only a 2% share (Média-métrie, see note 8). In our study conducted in spring 2009, everyone – mobile Internet users and non-users alike (some of them tried the mobile Internet and then gave up) – pointed out this computer habit. All evoked this primary use of the computer when they are at home, including Thomas (19 years old, his father has a restaurant, 12th grade, professional high school, smartphone Nokia, no flat rate mobile Internet but Wi-Fi connection), who said *"When I'm home, the computer is on, that's it."* How can this be explained? During the appropriation process of a technical device, the physical relationship to the object (i.e., holding it) allows us to discover it, get used to it and domesticate it in order to make it ours. Conjointly to this process, the mental and social representation of this incorporated relation builds itself. We think that the physical appropriation of the artifact for the mobile Internet remains a problem. The ergonomic dimension of the mobile phone plays an important role in the creation of the image of the mobile Internet. Almost everyone, even moderate users, expressed the problems related to factors such as the little size of the screen, pages too large to scroll across and the lack of comfort of the keyboard. This explains why even its users deviate from the prescribed use; they are adapting by setting up alternative and creative tactics which allows subversive uses (de Certeau, 1984): Thomas, for instance, has a Nokia specifically designed for video games so that he can play them anywhere at any time, but he nevertheless criticizes the ergonomics of the phone because *"regarding handiness, it's not a video game console, you know, because the keys of the mobile are too small"*. This makes us think that an adaptation process is at work. As Thomas explains, *"Anyways, when one likes it, ones has to adapt"*, which is the case when he is outside, although he admits linking it to the television as soon as he gets home *"and then I'm actually playing with the TV"*. There is also the case of William (19 years old, his stepfather is a mason, he lives alone in youth residence, 12th grade, professional high school, smartphone Samsung, flat rate mobile Internet), who is very proud of his Samsung touch screen. He explains that as soon as he gets home (at the youth residence actually, where he does not have Wi-Fi in his room), he uses his

mobile phone as a modem by linking it to his laptop, which he uses comfortably
with its keyboard and larger screen. We can qualify these two types of uses set
up by Thomas and William as a typical form of bypassing the usage prescribed
by the producers. It is a kind of tactics that the users set up to deal with the
producers' strategies (de Certeau, 1984).

A paradigmatic example is that of MSN. The instant messaging service
does not show any technical failure, but, even with the mobile version used in
2009, all of the interviewees who had tried MSN had given it up at the time of
the study, essentially for ergonomic or technical reasons. MSN is, by definition,
linked to an instant messaging image and this image is being challenged on the
cell phone. One has to scroll up into the conversation to see what the other has
written and, furthermore, *"it is really slow, by the time we type in an answer,
another question has already been asked"* explains Lila (18 years old, her father
is a truck stoker, 12th grade, professional high school, iPod Touch, no flat rate
mobile Internet, Wi-Fi Connection). Let us listen to what they think about their
experience. For Mona (20 years old, her father manages a small business, she
lives with her boy-friend, 12th grade, professional high school, not a smartphone
and only has the MSN) *"the constraints are the inconvenience"*. Likewise, Lila
explains: *"I don't like it that much [...] it's not practical"*. Benjamin (20 years
old, his father is an ex-cameraman retired, he lives with his girlfriend, 12th
grade, professional high school, iPhone with flat rate mobile Internet) says, *"It's
not convenient [...] we can't see the contacts, we can't see who we are talking
to"*. Nevertheless, he is very proud of his iPhone and describes the pleasurable
experience he lives when surfing on the Internet: *"it is comfortable to surf on the
Internet with the large screen of my iPhone"*. However, this screen becomes
"tiny" when Benjamin evokes his MSN uses. It clearly appears that the evaluation of the screen is not absolute; on the contrary, it is relative and the mobile
phone is compared to the personal computer for MSN use. Benjamin can be
described as an early adopter and a real opinion leader because he became very
enthusiastic about his iPhone and tried to convince us to adopt a new iPhone. To
sum up, the image of MSN on a mobile phone is slow, not practical, constraining
and inconvenient (this refers to the 2009 version of MSN). Furthermore, aside
from these technical and ergonomic elements, it seems that there is also a lack of
a more psychological dimension linked to the conversation itself, that is, to the
communication. William (male, 19 years old, his stepfather is a mason, he lives
alone in a youth residence, 12th grade, professional high school, smartphone
Samsung, flat rate mobile Internet) told us he does not use MSN on his mobile
phone: *"it's just plain useless, when you connect to MSN, it's for more than just
two minutes."* He would rather send a text message and then be online on MSN
in the evening. William explains that he must feel comfortable during the

conversation. More generally, we could add that some interviewees told us that the image they have of the Internet is still completely linked to the computer. For them, accessing "directly[22] the Internet" must be through a computer. This is consistent with the remark that the mobile Internet developed more rapidly in countries where personal computers are not as established as in Western countries. The European study that focused on mobile search (JRC, 2010) also noticed this point: "Though users generally continue to take computers as a reference point, respondents to our questionnaire indicated that they would consider mobile search as an alternative at home when the search experience becomes similar" (JRC, 2010, p. 11).[23] However, the issue of the alternative was still not mentioned by the non-user respondents to our questionnaire. Finally, the physical appropriation of the artifact and the social representation are mutually imbricated because they feed each other reciprocally.

5   Conclusion

We chose to focus on the image and the social representations of the mobile Internet to understand its emerging uses. The representation of the mobile Internet user was studied. This representation is in its construction, built by the apparatus of audience measurement and evolving into a consensual image. We are witnessing the constitution and construction of an audience, as defined by media reception studies and the mobile phone is actually becoming a mass medium due to the mobile Internet. We then analyzed the image built and produced in the public space by the marketing strategies that the network operators use to sell their mobile Internet offerings; the image is almost war-like, with aggressive marketing strategies and a certain symbolic violence. Then, we analyzed the social representation of this new media by both users and non-users. It initially appeared that the use of the mobile Internet remained limited among our sample. Second, the image of the computer still dominates access to the Internet for all the interviewees. It is not unheard of to think that the problems and barriers demonstrated by the irregular users of our sample are tactics and ways of using or ways of operating ["*manières de faire*"] (de Certeau, 1984)

---

[22]   The success of the iPhone is in part explained by the access to the "real" Internet it allows. It was the core of Steve Jobs's strategy (West & Mace, 2010). On the contrary, the operators' portals sites or the WAP model are maybe not completely foreign to this representation because they constituted walled gardens (Jaokar & Fish, 2006). This access restriction, completely opposed to the "spirit" of the image of the Web, could explain the failure of WAP in 2000. As a result, the walled gardens collapsed very rapidly (Mary Meeker, 2009).

[23]   Mary Meeker, Stanley Morgan's expert, forecasts that more users will likely connect to the Internet via mobile devices than desktop PCs within 5 years (Mary Meeker, 2009).

that, in the end, will allow the appropriation of this new media. It is important to be cautious about the images associated with infrequent use because they have to be replaced in their time and social context; they emerged in the social space at the time of our study (spring 2009) and this constitutes an ongoing and still-evolving process. It can be seen that the use of the mobile Internet is developing. The last quantitative study of the CRÉDOC (Bigot & Croutte 2009), carried out in June 2009, shows a certain increase in the use of the mobile Internet, the first since 2004: "The Internet on the mobile phone is finally taking off" (Bigot & Croutte, 2009, p. 43). However, if providers and economic experts have been eager to predict its uses since the start of the mobile Internet in 2004, sociologists are here to assert that the duration of the social formation of uses is a long one (Scardigli, 1992; Mallein & Toussaint, 1994; Perriault, 1989). We could also add that extreme personalization of uses is appearing. The discovery of the mobile Internet is often made through hybridization, or through a real combinatory of uses between personal computer, television and mobile phone. Indeed, in the study of the uses of the camera phone, we noticed how people set up a rationality of use to choose between the various devices available, e.g., between the mobile phone or the digital camera (Martin, 2009). The criteria of choice are composed of technical and economical constraints, as well as constraints related to the personal history of the subject and their previous cultural practices. However, we also noticed a discovery of amateur photography practices with the mobile phone by young people from lower class who had not previously owned a personal digital camera. This leads to another question: will the mobile Internet help people from the most unprivileged classes who do not have computers to access the Internet? The success of the iPhone must be more deeply analyzed because it does not concern only upper-class professionals; indeed, beyond its high price, one must also take into account the price of the monthly plan, which renders the annual cost substantial (Comparatively, it is interesting to notice that Apple computers are still limited to a certain well-off part of the population). What then will happen to the social distinction-based image that initiated its success? Other questions will appear, related to the permanent connectivity that the individual will have to face; checking emails all the time – also text messages sent from brands by mobile marketing – reading Facebook messages or producing personal content that can be spread on the Internet takes ineluctably part in the acceleration and the rise of urgency that the ICTs favored in our contemporary societies (Jauréguiberry, 2007). Will the hyper-connected individual develop tactics to manage all this, just as he learned to manage his reachability? (Martin, 2007; Licoppe & Levallois-Barth, 2009).

References

Bigot, R. & Croutte, P. (2009). *La diffusion des technologies de l'information et de la communication dans la société française*. CRÉDOC (Centre de recherche pour l'étude et l'observation des conditions de vie). Paris: CGTI/ARCEP. Available: http://www.arcep.fr/uploads/tx_gspublication/etude-credoc-2009-111209.pdf [November 30, 2009].

Bonnewitz, P. (2007). *Premières leçons sur la sociologie de P. Bourdieu*. Paris: Presses universitaires de France.

Bourdieu, P. (1984). *Distinction: A Social Critique of a Judgement of Taste*. Cambridge, MA: Routledge & Kogan Paul. (Original work published 1979)

Breton, P. & Proulx S. (2002). *L'explosion de la communication à l'aube du XXIe siècle*. Paris: Éd. La Découverte.

Caradec, V. (2001). Personnes âgées et objets technologiques: une perspective en termes de logique d'usage. *Revue française de sociologie, 42*(1), 117-148.

Certeau, M. de (1984). *The Practice of Everyday Life*. Berkeley: University of California Press. (Original work published 1980)

Donner, J. & Gitau, S. (2009). New Paths: Exploring Mobile-Centric Internet Use in South Africa. *Proceedings of the ICA Pre-Conference Mobile 2.0: Beyond Voice?* Chicago, Illinois, USA. Available: http://lirneasia.net/wp-content/uploads/2009/05/final-paper_donner_et_al.pdf [May 31, 2009].

Flichy, P. (2007a). *Understanding Technological Innovation: A Socio-Technical Approach*. Cheltenham: E. E. Publishing Ltd. (Original work published 1995)

Flichy, P. (2007b). *The Internet Imaginaire*. Cambridge, MA: MIT Press. (Original work published 2001)

Jaokar, A. & Fish, T. (2006). *Mobile Web 2.0: The innovator's guide to developing and marketing next generation wireless/mobile applications*. London: Futuretext.

Jauréguiberry, F. (2007). Les téléphones portables, outils du dédoublement et de la densification du temps: un diagnostic confirmé. *Tic&société, 1*(1). Available: http://ticetsociete.revues.org/281 [June 15, 2009].

Jodelet, D. (1991). *Madness and Social Representations: Living with the Mad in One French Community*. Berkeley: University of California Press. (Original work published 1989)

Joint Research Centre (JRC) of the European Commission and the Institute for Prospective Technological Studies (2010). Available: http://ec.europa.eu/dgs/jrc/index.cfm?id=1410&obj_id=10320&dt_code=NWS&lang=en [April 20, 2010].

Jouët, J. (2000). Retour critique sur la sociologie des usages. *Réseaux, 18*(100), 487-521.

Kapferer, J.- N. (2008). *The New Strategic Brand Management*. London: Kogan Page.

Katz, J. E. & Aakhus, M. (Eds) (2002). *Perpetual Contact: Mobile Communication, Private Talk, Public Performance*. Cambridge: Cambridge University Press.

Licoppe, C. & Levallois-Barth, C. (2009). Configurer l'accessibilité des voyageurs équipés à des services mobiles multimédia. *Réseaux, 27*(156), 15-48.

Ling, R. & Yttri, B. (2002). Hyper-coordination via mobile phones in Norway. In J. E. Katz and M. Aakhus (Eds.), *Perpetual Contact: Mobile Communication, Private Talk, Public Performance* (pp. 139-169). Cambridge: Cambridge University Press.

Ling, R. & Sundsoy, P. R. (2009). The iPhone and mobile access to the Internet. *Proceedings of the ICA Pre-Conference Mobile 2.0: Beyond Voice?* Chicago, Illinois, USA. Available: http://lirneasia.net/wp-content/uploads/2009/05/final-paper_ling_et_al. pdf [May 31, 2009].

Mallein, P. & Toussaint, Y. (1994). L'intégration sociale des technologies d'information et de communication: une sociologie des usages. *Technologies de l'information et société*, *6*(4), 315-335.

Martin, C. (2007). *Le téléphone portable et nous. En famille, entre amis, au travail.* Paris: L'Harmattan.

Martin, C. (2009). Camera phone and Photography among French young users. *Proceedings of the ICA Pre-Conference Mobile 2.0: Beyond Voice?* Chicago, Illinois, USA. Available: http://lirneasia.net/wp-content/uploads/2009/05/final-paper_martin.pdf [May 31, 2009].

Méadel, C. (2009). Publicité et mesure d'audience: La construction du plausible. In S. de Iulio (Ed.), *Savoirs et savoir-faire des professionnels de la publicité: Histoire et perspectives (1950-2009).* Strasbourg: Presses de l'Université de Strasbourg.

Méadel, C. (2010). *Quantifier le public: Histoire de la mesure d'audience de la radio et de la télévision.* Paris: Economica.

Méadel, C. & Bourdon, J. (2009). La boîte noire des mesures d'audience. Retour sur la réduction quantitative. *XVIe congrès de la Société Française des Sciences de l'Information et de la Communication (SFSIC),* Compiègne. Available: http://www.sfsic.org/congres_2008/spip.php?article110 [May 31, 2011].

Meeker, G. M. (2009). *The Mobile Internet Report.* Morgan Stanley Research. Available: http://www.morganstanley.com/institutional/techresearch/pdfs/mobile_internet_repo rt.pdf [May 31, 2011].

Minges, M. (2005). Is the Internet mobile? Measurements from the Asia-Pacific Region. *Telecommunications Policy*, *29*(2-3), 113-125.

Moliner, P. & Tafani, E. (1997). Attitudes and social representations: A theoretical and experimental approach. *European Journal of Social Psychology, 27* (6), 687-702.

Moscovici, S. (2001). *Social Representations: Explorations in Social Psychology.* Cambridge: Polity Press.

Pape, T. von & Martin, C. (2010). Non-usages du téléphone portable: au-delà d'une opposition binaire usagers/non-usagers. *Questions de communication, 18,* 113-144.

Perriault, J. (1989). *La logique de l'usage Essai sur les machines à communiquer.* Paris: Flammarion.

Scardigli, V. (1992). *Les sens de la technique.* Paris: Presses universitaires de France.

West, J. & Mace, M. (2010). Browsing as the killer app: Explaining the rapid success of Apple's iPhone. *Telecommunications Policy, 34*(5-6), 270-286.

Symbolic Models of Mobile Phone Appropriation
A Content Analysis of TV Serials

Veronika Karnowski

1   Introduction

Unintentionally overhearing a phone call while on a train, being advised to turn off one's cell phone in a cinema, or watching on the TV serials "Gilmore Girls" Lorelai Gilmore calling her daughter Rory on her mobile phone—these examples highlight the way in which mobile communication has become a part of our everyday lives. The mobile phone, originally designed solely as a telephone, is a cluster of innovations today. Consequently, to only analyze the adoption or rejection of these innovations is no longer sufficient: it is necessary to examine the process in which these innovations are integrated into the user's everyday life.

Images are involved in this process on two levels: First, a growing number of the services proposed for mobile phones are visual services, such as mobile television and the exchange of photographs. These services and their uses are considered in other articles within this volume. The present article focuses on a second level: as the innovation evolves, the way in which it is seen and represented (i.e., through audiovisual media) also changes. More precisely, this article traces the representation of the mobile phone within American and German TV serials between 1996 and 2006, during the first ten years of the widespread diffusion of the device.

These more symbolical aspects of an innovation's evolution are commonly neglected in empirical research on media innovation (i.e., in quantitative research), as such studies on media innovations tend to focus on their adoption and diffusion. A quantitative approach which is broad enough to consider symbolical questions is the "mobile phone appropriation model" (MPA-model) proposed by Wirth, von Pape and Karnowski (2008).

The central aspect of this model is meta-communication: i.e., communication about the innovation in question. The present study examines this meta-communication by integrating aspects from Social Learning Theory (Bandura, 1977) to identify the potential influence of mass media content on the individual appropriation process.

2   MPA-Model

The MPA-Model (cf. Wirth et al., 2008) deals with the question of how mobile phones are integrated in the user's daily routine. The model was developed on the basis of both adoption research (e.g., diffusion research, Theory of Planned Behavior (TPB), Technology Acceptance Model (TAM); see Ajzen, 2005; Davis, 1986; Fishbein & Ajzen, 1975; Rogers, 2003) and appropriation research (e.g., frame analysis, domestication, Uses-and-Gratifications approach; see Goffman, 1974; Katz, Blumler & Gurevitch, 1974; Silverstone & Haddon, 1996). Historically, it can be considered as an extension of the Theory of Planned Behavior (Ajzen, 2005) with the following four main characteristics (see figure 1):

1. The model considers appropriation to be a creative and active process, end-ing with patterns of individual usage and meaning. Thus, behavior is diffe-rentiated to its object-related and functional aspects. The object-related as-pects include fashion aspects (e.g., ring tones and accessories), handling as-pects and the general usage frequency of different functionalities such as te-lephony, text messaging or online services. The functional aspects represent the large variety of uses of the mobile telephone known from appropriation research and Uses-and-Gratifications studies (e.g., the management of eve-ryday life, maintaining relations; see Höflich & Rössler, 2001; Leung & Wei, 2000), with an emphasis on the symbolic dimension (e.g., status; see Leung & Wei, 2000) of this process.
2. The model takes into account the symbolic value of the object mobile tele-phone and its usage.
3. The model no longer takes TPB's independent variables, "behavioral be-liefs", "normative beliefs" and "control beliefs" as static, but understands them as the constantly evolving results of the appropriation process (Jonas & Doll, 1996; Kendzierski, 1990). Consequently, the model is conceptua-lized as a cycle, with appropriation being a constantly renewed process. Pragmatic and symbolic use is not only the result of behavioral, normative and control beliefs, but also their basis (Oulette & Wood, 1998).
4. The impact of communication on the appropriation process: meta-communication. Thus, behavioral, normative and control beliefs, as well as symbolical and practical behavior, are negotiated through communication among users, with producers and mass media, be it mass communication, personal influence or the simple demonstration and observation of one's mobile phone use.

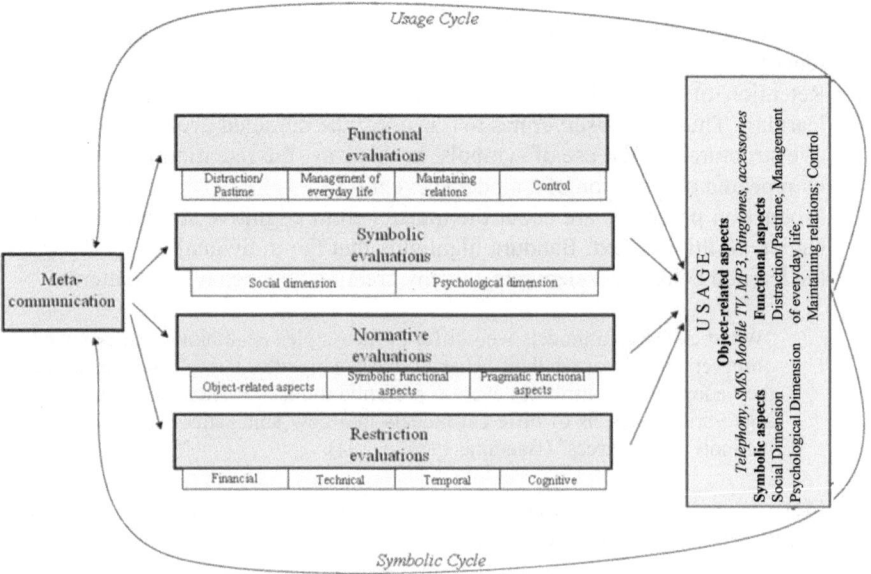

Figure 1: The MPA-Model (Wirth et al., 2008, p. 606)

The concept of meta-communication already implies the influence of mass communication on the individual appropriation process, without amplifying this idea. The present paper will show how Bandura's (1977) Social Learning Theory can be integrated into this concept of meta-communication.

3   Social Learning Theory

Social Learning Theory is based on the idea of vicarious learning by observing models supported by different motivational aspects. Bandura (1977, 1986) describes four sub-processes of observational learning (see figure 2):

1.   Attentional processes manage the selection of modeled events. This process is influenced by salience, affective valence, complexity, prevalence, and the functional value[1] of the modeled event, as well as the observer's perceptual

---

[1]   *Salience* refers to the degree to which a modeled behavior stands out among other behaviors, *affective valence* refers to whether this behavior is seen as 'good' or 'bad', *prevalence* refers to

set and capabilities, cognitive capabilities, arousal level, and acquired preferences.

2. Retention of modeled events is the second precondition for observational learning. Thus, the observer has to transform the modeled events into cognitive structures by the use of symbols. In this way, the retention is facilitated by repeated observation of a modeled event.

3. Production processes are about the transfer from cognitive structures to behavior. In this context, Bandura highlights that the individual can rearrange the learned behavioral elements, thereby creating new behavioral patterns:

> "When exposed to models who differ in their styles of thinking and behavior, observers rarely pattern their behavior exclusively after a single source, nor do they adopt all the attributes even of preferred models. Rather, observers combine various aspects of different models into new amalgams that differ from the individual sources" (Bandura, 1986, p. 104).

4. Motivational processes determine those observed events which are imitated. These motivational aspects can be external incentives as well as vicarious incentives; i.e., observing the modeled behavior being rewarded. In addition, the individual can reward herself/himself (self-incentives) by anticipating positive outcomes of the behavior in question.

According to Bandura (1986), the modeled events may be either observed in the direct surroundings of an individual or in the mass media, without any fundamental difference regarding effects. The latter is called "symbolic modeling". Even though Bandura (1986) does not define the term "media", it can be argued that he acts on the assumption of mass media, as he defines symbolic models by their transmission to a large number of recipients:

> "[...] it can transmit simultaneously knowledge of wide applicability to vast numbers of people through the medium of symbolic models." (Bandura, 1986, p. 47)

---

how often the behavior is observed, and *functional value* describes whether the behavior is useful for the individual.

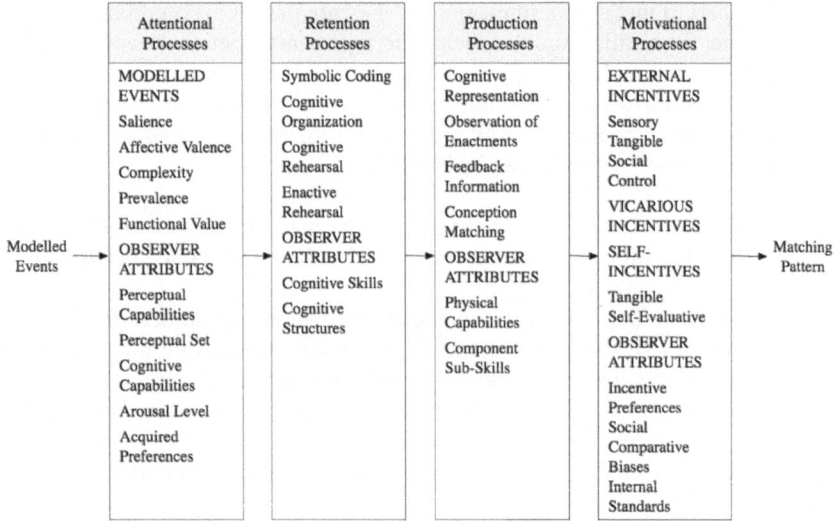

Attentional Processes	Retention Processes	Production Processes	Motivational Processes
MODELLED EVENTS	Symbolic Coding	Cognitive Representation	EXTERNAL INCENTIVES
Salience	Cognitive Organization	Observation of Enactments	Sensory
Affective Valence	Cognitive Rehearsal		Tangible
Complexity		Feedback Information	Social Control
Prevalence	Enactive Rehearsal	Conception Matching	VICARIOUS INCENTIVES
Functional Value			
OBSERVER ATTRIBUTES	OBSERVER ATTRIBUTES	OBSERVER ATTRIBUTES	SELF-INCENTIVES
Perceptual Capabilities	Cognitive Skills	Physical Capabilities	Tangible Self-Evaluative
Perceptual Set	Cognitive Structures	Component Sub-Skills	OBSERVER ATTRIBUTES
Cognitive Capabilities			Incentive Preferences
Arousal Level			Social Comparative
Acquired Preferences			Biases
			Internal Standards

Modelled Events → ... → Matching Pattern

Figure 2: Four sub-processes of observational learning (Bandura, 1986, p. 52)

In addition, Bandura (1986, pp. 55, 70, 145, 166, 318, 511) explicitly and repeatedly refers to TV as a mediator of symbolic models. Accordingly, both the MPA-Model and Social Learning Theory assume fictional characters in the mass media who use mobile phones to influence the individual appropriation process of mobile phone users in real life. To examine this influence, it is necessary to undertake empirical research firsthand on these symbolic models of mobile phone appropriation displayed in the mass media. Consequently, the aim of the present study is to identify and describe fictional symbolic models of mobile phone appropriation displayed on TV. The analysis is restricted to symbolic models in family TV serials which concentrate on everyday life topics, because two aspects crucial to the imitation of modeled events are fostered in this type of TV serial: the modeled events are highly probable to be observable repeatedly and the behavior displayed is close to that in real life.

4   Symbolic Models of Mobile Phone Appropriation

According to the MPA-Model (Wirth et al., 2008), symbolic models of mobile phone appropriation (i.e., the use of mobile phones as displayed in the mass media) are part of meta-communication, as they are part of mass communication. At the same time, this modeled behavior represents (fictional) appropriation processes which can by described using the MPA-Model. Consequently, the MPA-Model is mirrored in the dimension of meta-communication, in order to describe symbolic models of mobile phone appropriation. Based on this idea, two levels of the MPA-Model can be identified:

- Level 1: Mobile phone appropriation by real-life individuals (MPA I)
- Level 2: Mobile phone appropriation by fictional characters (MPA II)

Thus, symbolic models of mobile phone appropriation can be described by the different elements of the MPA-Model. Theoretically, this is true for all parts of the MPA-model. As we only observe the behavior of fictional protagonists, and because we have no insight into their (fictional) cognitive structures, only the behavioral dimensions of the MPA-model, as well as the aspect of meta-communication (in meta-communication), are relevant in describing second-level appropriation processes.

In addition, symbolic models can be described in terms of motivational aspects according to Social Learning Theory. These motivational aspects are an important influence on the probability of mobile phone users imitating the modeled behavior (see above).

5   Research Questions

The present paper aims to examine the potential influence of mass media content on the individual appropriation process. Accordingly, it concentrates on the period since the end of the 1990s, when mobile phone usage in Western societies showed a sudden leap in terms of both penetration and total numbers (see Agar, 2003; CTIA, 2007): Which symbolic models of mobile phone appropriation can be identified in family TV serials? How did they evolve during the period from 1996 to 2006?

6   Methodology

To address the above questions, content analysis was performed for five family TV serials for which more than 80 episodes were first aired in Germany between 1996 and 2006: *Lindenstrasse* (produced by Hans W. Geißendörfer, 1985–), *Dawson's Creek* (produced by Gregory Prange, 1998–2003), *Sex and the City* (produced by Michael P. King, 1998–2004), *Gilmore Girls* (produced by Gavin Polone and Amy Sherman, 2000–2007) and *O.C. California* (produced by Josh Schwartz, 2003–2007). *Lindenstrasse* is a German production, whereas all the other serials are US productions.[2]

As a first step, all of the episodes of these TV serials (944 episodes in total) were screened. As a second step, content analysis was performed for all scenes in which a mobile phone could be seen or heard, or was the subject of a dialogue. A scene was defined as follows:

• The end of a scene is characterized by a simultaneous change of topic and protagonists or a simple change of topic, when the new topic is covered for more than 30 seconds consecutively. Short insertions in the dialogue do not terminate a scene provided they are shorter than 30 seconds. If events jump from one strand of a plot to another and back, both fragments of a scene are analyzed altogether as one scene.
• A protagonist is any person in a scene who is talking or nonverbally commenting on events. The protagonists in a scene do not have to be located in the same place, and they can be connected by any means of communication.

In addition to formal aspects such as duration, serial title, season and episode, the behavioral aspects of the modeled behavior were analyzed according to the MPA-model (see above), including

1. the object-oriented usage aspect: mode of usage (calling, text messaging, etc.), design (chin-chins, logos, color of mobile device, etc.), ring tone, handling;
2. the functional usage aspect: distraction/pastime, management of everyday life (coordination and exchange of information), maintaining relations, control.

In addition to these MPA-aspects, three motivational aspects were coded based on Bandura's (1977) Social Learning Theory:

---

2   Thirteen TV serials satisfied these criteria. Most were German (seven serials) or US productions (four serials). A typical German TV serial (*Lindenstrasse*) was chosen, which was aired throughout the entire timeframe. As none of the US serials was aired throughout the entire timeframe, all four US serials were analyzed, resulting in a sample of 521 episodes of a German production and 413 episodes of US productions. Clearly, this sample cannot be considered representative; consequently, it only provides an initial insight into symbolic models of mobile phone appropriation offered by TV serials.

1. the functional value of a modeled behavior (i.e. the success of a protagonist in attaining his/her actual goals);
2. the probable similarity between modeled behavior and recipient (gender and phase of life of the protagonist);
3. vicarious reinforcement (i.e., commendation or criticism in meta-communication[3]).

The content analysis was conducted between February and July 2007. The reliability of the variables was between 0.83 and 1.00 (average 0.96).

7   Data Analysis

To identify the outcomes of first-level appropriation processes, Wirth et al. (2008) suggest using clustering techniques. Consequently, the clustering technique of latent class analysis (LCA) was employed here, which has the following advantages over traditional cluster analysis:

1. LCA allows for classifying variables of each level of measurement: even different levels of measurement can be integrated in the analysis. This is especially helpful for clustering content-analysis data (see also Matthes, 2007).
2. LCA does not necessarily result in a cluster solution: it can reject clustering of the data (e.g., Fraley & Raftery, 1998).
3. LCA provides statistical tests with which to identify the exact number of clusters; consequently, it is less arbitrary than traditional cluster analysis.
4. By virtue of its probabilistic conception, LCA takes into account that the clustered variables may be neither completely reliable nor completely valid.

8   Results

In total, 1,413 symbolic models of mobile phone appropriation were identified, including the usage of mobile phones as well as mobile phones simply being visible in the scene. Although the material analyzed consisted of 521 episodes of *Lindenstrasse* and 413 episodes of US serials, symbolic models of mobile phone appropriation were much more commonly found in the US serials (82% of the total cases; see table 1).

---

[3]   Other aspects of meta-communication have also been coded, but are not analyzed in this paper.

Table 1: Symbolic models by serials

	Frequency (N = 1,413)	Symbolic models per episode
Lindenstrasse	18%	0.5
US serials	82%	2.8

In episodes from the early part of the study period, few symbolic models of mobile phone appropriation could be observed, with an increase only since 2003, lagging behind uptake of the device in real life (see figure 3).

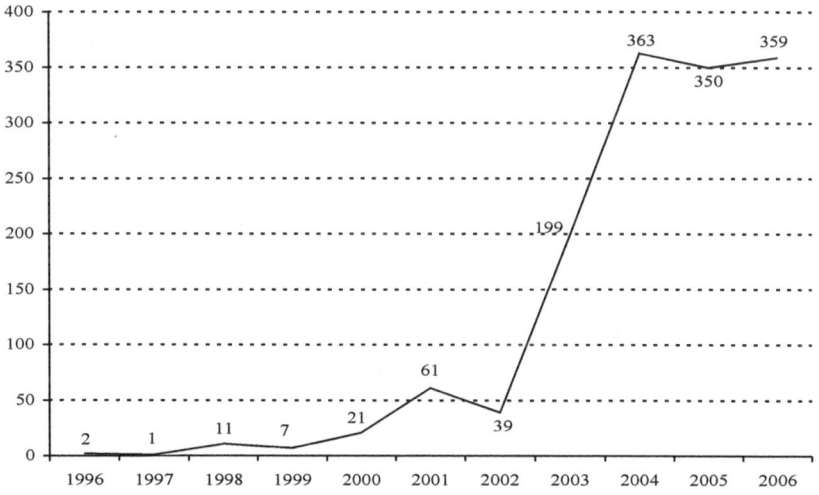

Figure 3: Annual number of symbolic models of mobile phone appropriation in TV serials

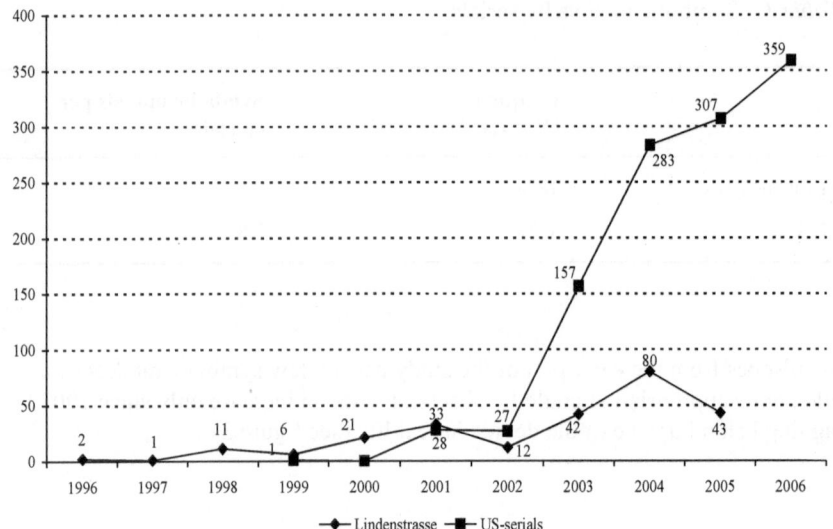

Figure 4: Annual number of symbolic models of mobile phone usage in
Lindenstrasse versus US TV serials

This enormous increase in 2003 is due mainly to the US serials, in which the
number of symbolic models of mobile phone appropriation decupled between
2002 and 2004. In *Lindenstrasse*, the number of symbolic models of mobile
phone appropriation increased only slowly (see figure 4).

Clustering

Because the clustering of second-level appropriation patterns is based on all the
functional and object-oriented usage aspects, only the 846 symbolic models of
fictional characters using a mobile phone are integrated in the analysis. Since the
two object-oriented usage dimensions of mobile phone accessories (chin-chins
and logos) are nearly constant (they are observed in only 1% of the symbolic
models analyzed), they are not integrated in LCA.

       To identify the number of clusters, the one- to ten-cluster solutions were
calculated and compared. All solutions have a non-significant p-value (likelihood-ratio test); thus, the model prediction does not differ significantly from the
observed data. The Cressie–Read test and Pearson's χ^2 yield a significant p-value

for the one-cluster solution, which can therefore be eliminated. To further test the validity of the results, the likelihood-ratio test was checked by bootstrapping. This method demonstrated that the two- and three-cluster solutions differ significantly from the observed data.

Table 2: Likelihood ratio (including bootstrapping), Cressie–Read, Pearson's χ^2 and Bayesian Information Criterion (BIC) for one- to ten-cluster solutions

	Likelihood Ratio		Cressie–Read	Pearson's χ^2		BIC
	p-value	p-value (Bootstrap)	p-value		p-value	
1-Cluster	739	**0.99**	0.00	2038	0.00	7999	0.00	7350
2-Cluster	525	**1.00**	0.00	563	**1.00**	686	**1.00**	7223
3-Cluster	338	**1.00**	0.00	408	**1.00**	579	**1.00**	7124
4-Cluster	**266**	**1.00**	**0.17**	**316**	**1.00**	**441**	**1.00**	**7140**
5-Cluster	238	**1.00**	**0.32**	286	**1.00**	405	**1.00**	7200
6-Cluster	202	**1.00**	**0.68**	236	**1.00**	308	**1.00**	7251
7-Cluster	175	**1.00**	**0.75**	184	**1.00**	219	**1.00**	7312
8-Cluster	158	**1.00**	**0.83**	184	**1.00**	235	**1.00**	7382
9-Cluster	148	**1.00**	**0.58**	158	**1.00**	189	**1.00**	7460
10-Cluster	131	**1.00**	**0.76**	138	**1.00**	166	**1.00**	7530

Thus, only the four- to ten-cluster solutions are considered (see table 2). Generally, the most suitable solution is that with the fewest parameters to be estimated, thereby having the lowest BIC value. In the present case, this corresponds to the four-cluster solution.

Usage clusters

LCA yields the specific probabilities of different parameter values integrated in the analysis for each cluster. The different clusters can be described based on these probabilities (see table 3).

Table 3: Average probabilities of parameter values, explained variance of classified variables and relative cluster size of the four-cluster solution

		Cluster 1	Cluster 2	Cluster 3	Cluster 4	R^2
Functional usage aspects	Distraction/ pastime	1%	0%	6%	23%	8%
	Management of everyday life: coordination	**52%**	**99%**	1%	18%	61%
	Management of everyday life: information exchange	**99%**	0%	12%	24%	81%
	Maintaining relations	4%	28%	**63%**	4%	28%
	Control	3%	4%	10%	3%	2%
Object-oriented usage aspects	Usage mode: telephony	**100%**	**100%**	**100%**	36%	60%
	Usage mode: other	0%	1%	2%	**99%**	81%
	Color of mobile phone — grey, silver, black	**93%**	**82%**	**86%**	**75%**	2%
	Color of mobile phone — other colors	5%	15%	11%	23%	
	Color of mobile phone — not visible	1%	3%	3%	3%	
	Handling — visible	14%	16%	19%	21%	1%
	Handling — not observable	27%	37%	29%	32%	
	Handling — other	7%	3%	5%	3%	
	Handling — not visible	51%	44%	47%	44%	
Relative size		34%	31%	30%	5%	

Cluster 1: Management of everyday life by telephone

Rory calls her mum Lorelai using her silver mobile phone. She asks her where to take her car for a service. Lorelai tells her to ask her grandfather for advice. They arrange to meet at the grandparents for dinner on Friday night.

This hypothetical scenario is typical of the usage situations in this cluster. The purpose of mobile phone usage is generally the management of everyday life. The cell phone is displayed as a discreet object. The color of the device is most likely to be grey, silver or black. The probability that the mobile phone is visible to others before being used is lowest in this cluster.

Cluster 2: Coordination by telephone

Carrie reaches Miranda on her pink mobile. They agree to have lunch together on the following day.

The purpose of this conversation is coordination. Although the colors of the mobile devices in this cluster are mainly grey, silver or black, the probability of colored devices (15%) is higher than that in clusters 1 and 3.

Cluster 3: Maintaining relations by telephone

Gail calls her son Dawson who lives far away in California. She first wants to know what he's doing right now. Then Dawson tells her about a party he went to yesterday and about a girl he met there.

Here, the mobile is used most likely to maintain relations. The aspect of control is more likely in this cluster than in the others. As with all the other clusters, the mobile phones are mainly grey, silver or black, and are carried discreetly and are not visible to the other fictional characters.

Cluster 4: Usage of other functionalities

Gabi sits at the station waiting for her train. She takes her blue mobile out of her bag and sends a text message to her husband, telling him that her train is late.

Functionalities other than telephony—especially text messaging—are used. The most likely purpose of usage is as a pastime or management of everyday life. The probability of colored mobiles is higher in this cluster than in clusters 1 to 3. This cluster also comprises the use of visual services on the mobile phone, such as taking photos or watching videos, which is rarely observed in TV serials. Such services were used in only 4 of the 846 analyzed usage situations.

Motivational Aspects

We now examine the motivational aspects of the identified usage patterns, which are crucial for a possible imitation of these behavioral patterns by the recipients. To this end, each case is attributed to the cluster to which it most likely belongs.

The classification error (i.e., the proportion of cases which are incorrectly classified) is 3.6%.

Similarity

Men organize their everyday lives by telephone to a significantly greater degree than do women, whereas all other usage patterns are more commonly conducted by female characters. In particular, the usage pattern of other functionalities is dominated by women. Thus, TV serials clearly show stereotypical differences in usage patterns for male and female protagonists (see table 4).

Table 4: Gender

	Male	**Female**
Management of everyday life by telephone (n = 267)	65%	35%
Coordination by telephone (n = 242)	45%	55%
Maintaining relations by telephone (n = 238)	37%	63%
Use of other functionalities (n = 33)	30%	70%
Total (n = 780)	49%	51%

Pearson's χ^2 = 46.7 (p < 0.001)

Table 5: Phase of life

	Teenagers	Adults	Elderly adults
Management of everyday life by telephone (n = 267)	42%	55%	3%
Coordination by telephone (n = 242)	48%	50%	2%
Maintaining relations by telephone (n = 238)	55%	43%	3%
Usage of other functionalities (n = 33)	48%	52%	0%
Total (n = 780)	48%	50%	3%

Pearson's χ^2 = 10.1 (n.s.)

There are no significant differences in usage patterns with regard to the phase of life: all usage patterns are accomplished by both teenage and adult protagonists. Symbolic models of mobile phone appropriation by elderly adults are rarely observed (see table 5).

Success

Most of the mobile phone usage shown in TV serials is successful. The usage patterns for management of everyday life by telephone and coordination by telephone are significantly more successful than the other two usage patterns (see table 6). Thus, imitation of the usage pattern for the management of everyday life is most probable.

Table 6: Success

	Successful	Ambi-valent	Not successful	Unknown
Management of everyday life by telephone (n = 289)	68%	11%	13%	8%
Coordination by telephone (n = 265)	71%	9%	11%	9%
Maintaining relations by telephone (n = 254)	52%	10%	19%	19%
Usage of other functionalities (n = 39)	59%	3%	10%	28%
Total (n = 847)	64%	10%	14%	12%

Pearson's χ^2 = 43.2 (p < 0.001)

Commendation vs. Criticism of Usage Patterns

Explicit commendation or criticism of usage patterns via the comments of other fictional characters is rarely observed, although criticism is dominant over commendation. This is especially true for the usage of other functionalities. Coordination by telephone is most often reinforced, although it is still criticized more often than commended (see table 7). Considering this motivational aspect, only a minor influence of the symbolic models of mobile phone appropriation is to be supposed.

Distribution of Usage Clusters among German and US Serials

While all three telephony usage clusters are equally observable in the US serials, the management of everyday life dominates in *Lindenstrasse*. Maintaining relations by telephone occurs significantly less frequently in *Lindenstrasse* than in the US serials, whereas the usage of other functions is more common in *Lindenstrasse* (see table 8).

Table 7: Commendation vs. criticism of usage patterns

	Commen-dation	Ambi-valent	Criticism	No commenda-tion or criticism
Management of everyday life by telephone (n = 69)	0%	3%	28%	70%
Coordination by telephone (n = 44)	11%	0%	16%	73%
Maintaining relations by telephone (n = 51)	4%	4%	25%	67%
Usage of other functionalities (n = 10)	0%	20%	20%	60%
Total (n = 174)	4%	3%	24%	69%

The X^2-test is not applicable, as 56% of the cells in the table have an estimated frequency of less than 5.

Table 8: Usage clusters by serials

	Lindenstrasse (n = 129)	US serials (n = 717)	Total (n= 846)
Management of everyday life by telephone	41%	33%	34%
Coordination by telephone	33%	31%	31%
Maintaining relations by telephone	18%	32%	30%
Usage of other functionalities	8%	4%	5%

Pearson's χ^2 = 14.9 (p < 0.01)

Distribution of Usage Clusters over Time

Because of the low number of cases in the early years of the study period, temporal trends are only considered for the period since 2000.

Management of everyday life by telephone, maintaining relations by telephone and coordination by telephone each make up approximately one third of usage patterns since 2000. The rate of maintaining relations by telephone was largely constant until 2006, whereas the rate of management of everyday life shows a slight decrease. Usage of other functionalities was only observed since 2003 and shows a low rate in all years (see figure 5).

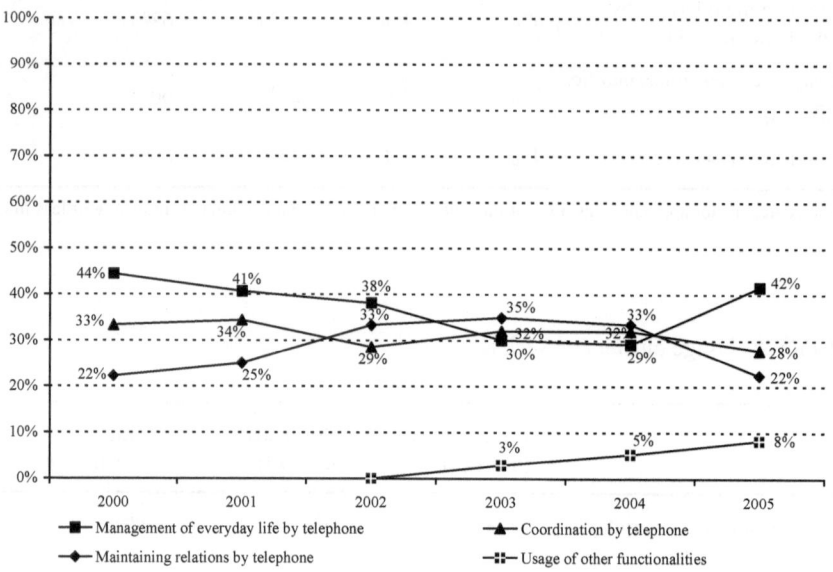

Figure 5: Usage clusters over time

Because of the small sample size, there exist large up- and downturns in the occurrence of the different usage clusters in *Lindenstrasse*. Between 2000 and 2002, management of everyday life was dominant, the contribution of maintaining relations by telephone showed a decline, and coordination by telephone consistently made up around one third of the observed usage clusters. Since 2004, the situation has changed due to a constant increase in usage of other functions, whereas the management of everyday life by telephone has shown a downward trend (see figure 6).

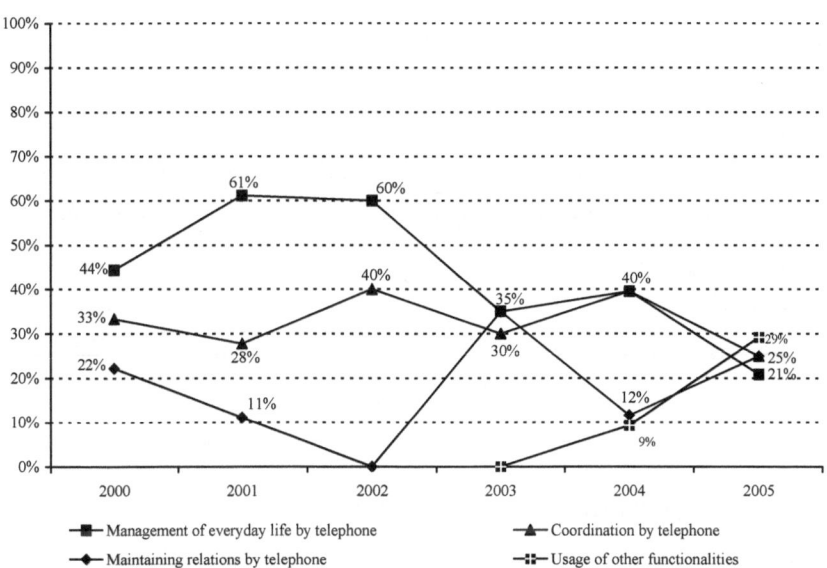

Figure 6: Usage clusters over time (*Lindenstrasse*)

Considering US serials, the situation is slightly different. Taken together, each of the three telephony-usage clusters makes up approximately one third of the usage situations. The contribution of maintaining relations by telephone shows a slight decline, whereas management of everyday life shows a slight increase. Coordination by telephone remained stable at around one third. Usage of other functionalities first appears in 2003, but does not exceed 10% (see figure 7).

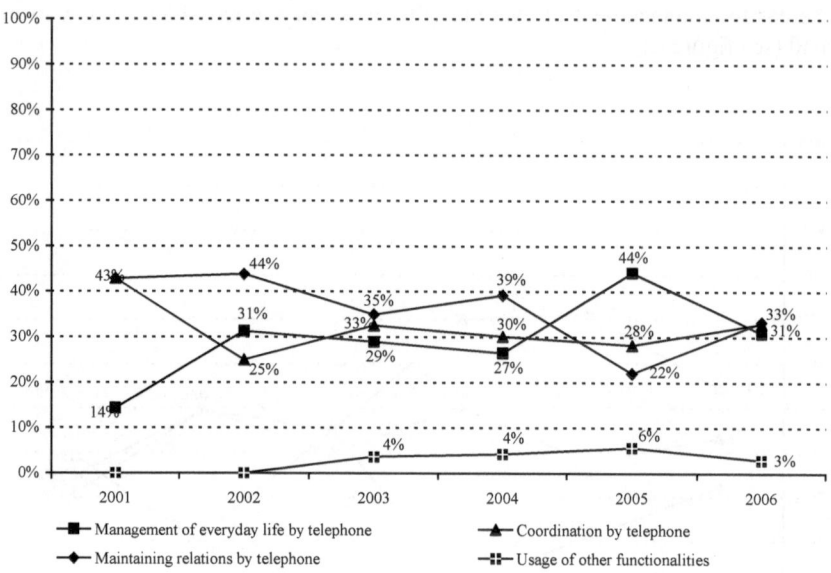

Figure 7:   Usage clusters over time (US serials)

9   Summary

As with the usage patterns which can be identified as outcomes of a first-level appropriation process (see Wirth et al., 2008), LCA could identify second-level usage patterns: management of everyday life by telephone, coordination by telephone, maintaining relations by telephone and usage of other functionalities. Thus, mirroring the MPA-model in the dimension of meta-communication is suitable and useful in order to describe aspects of meta-communication.

Furthermore, it was shown that the various usage patterns differ mainly in terms of the functional usage aspects, as well as in the mobile phone services used. Contrary to the first-level usage patterns identified by Wirth et al. (2008,), which clearly differ from each other in terms of the object-oriented usage aspect, handling and design are of little importance to the identified second-level appro- priation patterns: mobile phones in TV serials are consistently displayed dis- cretely in terms of handling and design.

All three telephony-usage clusters were observed in both US serials and *Lindenstrasse.* While the different clusters make similar contributions in the US serials, the usage cluster of management of everyday life is dominant in *Lindenstrasse*. Usage of other functionalities is first seen in 2003, but makes only a minor contribution in all subsequent years.

Examining the motivational factors associated with the usage clusters, we note some interesting differences. First, the assignment of usage clusters to male and female actors is based mainly on stereotypes. The naturally brief, factor-oriented phone calls in the management of everyday life are conducted significantly more often by male protagonists, whereas female actors dominate the human relations usage pattern of maintaining relations. Thus, it can be supposed that the symbolic models displayed in the serials also support the recipients' stereotypical usage patterns. Second, the usage patterns of management of everyday life (management of everyday life by telephone and coordination by telephone) are the most successful usage patterns in the TV serials examined. Thus, imitation of theses usage clusters is most likely.

Taken together, the symbolic models of mobile phone appropriation generally provide positive behavioral models for the individual appropriation process in terms of an adequate usage of this innovation. These models represented in audio-visual media have the potential to influence the uses of mobile communication themselves, as they literally show the recipients how to use a mobile phone and which symbolic value to attribute to it.

It must be remembered that the present study could only throw a first glance at the dimension of meta-communication in the individual appropriation process. To assure the validity of the present findings, further content analysis is necessary, also covering non-fictional content. Based on such findings, it would be possible to connect data on meta-communication with findings on individual appropriation patterns in real life as a second step, in order to analyze the influence of meta-communication (i.e. interpersonal and mass communication) on the individual appropriation process.

References

Agar, J. (2003). *Constant Touch: A Global History of the Mobile Phone*. Cambridge: Icon Books.

Ajzen, I. (2005). *Attitudes, Personality, and Behavior*. Milton-Keynes: Open University Press/McGraw-Hill.

Bandura, A. (1986). *Social Foundations of Thought and Action. A Social Cognitive Theory*. Upper Saddle River: Prentice Hall.

Bandura, A. (1977). *Social Learning Theory*. Englewood Cliffs: Prentice-Hall.

CTIA (2007). *Wireless quick facts: Mid-year figures 2007*. Available: http:// www.ctia. org/advocacy/research/index.cfm?bPrint=1&AID=10323 [December 12, 2007].

Davis, F. D. (1986). *A Technology Acceptance Model for Empirically Testing New End-User Information Systems: Theory and results*. Boston: MIT.

Fishbein, M. & Ajzen, I. (1975). *Believe, Attitude, Intention and Behavior: An Introduction to Theory and Research*. Reading: Addison-Wesley.

Fraley, C. & Raftery, A. (1998). How many clusters? Which clustering method? Answers via model-based cluster analysis. *The Computer Journal, 41*(8), 578-588.

Goffman, E. (1974). *Frame Analysis: An Essay on the Organization of Experience*. Cambridge: Harvard University Press.

Höflich, J. R. & Rössler, P. (2001). Mobile schriftliche Kommunikation oder: E-Mail für das Handy. *Medien & Kommunikationswissenschaft, 49*(4), 437-461.

Jonas, K. & Doll, J. (1996). Eine kritische Bewertung der Theorie überlegten Handelns und der Theorie geplanten Verhaltens. *Zeitschrift für Sozialpsychologie, 27*(1), 18-31.

Katz, E., Blumler, J. G., & Gurevitch, M. (1974). Utilization of mass communication by the individual. In J. G. Blumler & E. Katz (Eds.), *The Uses of Mass Communications. Current Perspectives in Gratifications Research* (pp. 249-168). Beverly Hills: Sage Publications.

Kendzierski (1990). Decision making versus decision implementation: An action-control approach to exercise adoption and adherence. *Journal of Applied Social Psychology, 20*(1), 27-45.

Leung, L. & Wei, R. (2000). More than just talk on the move: Uses and gratifications of the cellular phone. *Journalism and Mass Media Quarterly, 77*(2), 308-320.

Matthes, J. (2007). *Framing-Effekte. Zum Einfluss der Politikberichterstattung auf die Einstellungen der Rezipienten*. München: Reinhard Fischer.

Oulette, J. A. & Wood, W. (1998). Habit and Intention in Everyday Life: The Multiple Processes by Which Past Behavior Predicts Future Behavior. *Psychological Bulletin, 124*(1), 54-74.

Rogers, E. M. (2003). *Diffusion of Innovations*. New York: Free Press.

Silverstone, R. & Haddon, L. (1996). Design and the domestication of information and communication technologies: Technical change and everyday life. In R. Silverstone & R. Mansell (Eds.), *Communication by Design. The Politics of Information and Communication Technologies* (pp. 44-74). Oxford: Oxford University Press.

Wirth, W., von Pape, T. & Karnowski, V. (2008). An integrative model of mobile phone appropriation. *Journal of Computer-mediated Communication, 13*(3), 593-617.

The Image of Youth in Mobile Phone Advertising

Miguel Ángel Nicolás Ojeda

1 Introduction

This work analyzes the graphical advertisements of mobile telephone companies in Spain. More specifically, we study the representation of young people in advertising. The methodology, following the linguistic tradition, is a social and semiotic analysis of ads; this includes an analysis of the expression and content in samples of ads that have been gathered from various newspapers.

The results offer us quantitative information regarding the principal persuasive arguments of the sector, the products, the social information of the prominent characters in ads and the profiles or groups of images of young people associated with mobile telephones.

In this work, we study the creativity and business aims in mobile advertisements and their relationship to the images of young people used in the ads to determine their roles within the business strategy of the campaign. We focused on data from 2009 but started with the data collected in our 2006 doctoral thesis, *Publicidad y Juventud: Un análisis sociosemiótico*, in which we included a study that was parallel to this study and comprised six years (2000 to 2005).

2 Context: Consumer Society, Young Consumers and Techno-Culture.

The current coexistence model of capitalist Western societies leads their citizens to participate in a continuous and blindingly speedy transformation of their consumerist lifestyles that is influenced by the continuing evolution of industry, technology and the application of such technology. Our approach accepts a state of change because not doing so would deny the obvious. However, it is also equally obvious that the consumer society, from its inception, has always been constituted by consumers and is always changing its semantic relation to goods and services. "Consumer capitalism is not automatically born with manufacturing techniques capable of producing standardized goods in large series. It is also a cultural and social construction that equally requires the education of consumers and the visionary spirit of creative entrepreneurs, the invisible hand of managers" (Lipovetsky, 2007, p. 26, translation by author).

This self-awareness of change is more evident among consumers now than perhaps it might have been in other phases because change, even though it is conditioned by technological diversity, emerges from consumers in a conscious and intentional way. In his book *Le Bonheur Paradoxal* (The Paradoxical Happiness), Lipovetsky lists the three ages of consumer capitalism and argues that society has entered a new phase of *"hyper-modernity"*, characterized by the hypermodern individual, and *"hyper-consumption"*, which absorbs and integrates into an increasing number of spheres of social life and encourages individuals to consume for their own personal pleasure rather than to enhance their social status. These three states allow us to address the understanding of consumer society as an evolving state, able to influence and be the main propeller in the constitution of our daily lives. According to Lipovetsky, we have moved into a third phase of our consumer society, whose individual consumer was forged in the previous phase (60s) through household commodities (telephone, television, a family car). The current phase (III) is conditioned by omnipresent hyper-consumption. "The society of hyper-consumption can write on their banners, with triumphant words: 'To each its objects, to each its use, to each its lifestyle'" (Lipovetsky, 2007, p. 97). Consumption is hyper-individualized, professionalized and stretches over time. The reflections of Lipovetsky's essay lead us to envisage a society of continual use, free from time-space frameworks and, of course, a cyber-consumer. For Lipovetsky and many marketing strategists, signs no longer have the same spatial power in commercial relationships, but they now have a temporal power. "The effort to compress time has been understood as one of the signs of advent of a new temporary condition of man, characterized by sacralization of present, for an 'Absolute present', self-sufficient and increasingly detached from past and future" (Lipovetsky, 2007, p. 105).

However, according to Lipovetsky, the *turbo-consumer* reflects a series of individual strategies. Thus, he participates, as a direct agent, in shaping the society in which he lives and needs radical shifts in the use of his time to search for welfare. Our brief approach to Lipovetsky's work enables us to understand that the consumer society is changing and is influenced by multiple factors. We can distinguish the relationships that provide consumers with products and brands, the motivations that lead to their consumption and the role of advertising as a mediator for that relationship.

The recent work of Victor Gil and Felipe Romero, *Crossumer*, offers an enriching vision through which to conceptualize and name this new consumer, the relationships he maintains with advertising, his use of Web 2.0 and mobile technologies, his location and the ways to research him. According to Gil and

Romero, the term "Crossumer" describes a new consumer configured by his distrust of brand communication, who knows the marketing strategies and intentions and who actively participates in the acceptance and rejection of brand messages (2008, p. 27). This consumer emerges in a context of technological development and omnipresent advertising. He is presented as a consumer with power, able to modify and influence the development of advertising campaigns, validate their content, act as an intermediary and yet not be influenced (or not believing to be influenced) by their messages. He uses a social web to build his own individualized community created from his own tastes, and above all, he wants to buy at the lowest price. The mobile phone is one of new consumer's top players, and young people have been fundamental in its technological evolution and social use. "Young people are among the major consumers of mobile phone technology and are often considered to be forerunners in its adoption and evolution" (Thulin & Vilhelmonson, 2008, p. 138). Understanding the new consumer necessarily requires awareness of the uses and functions that young people attach to the mobile phone and how it has become one of their most attractive technologies. Richard Ling reminds us that "many adults have the sense that the use of mobile telephones among teens is a whole different world. They experience teens as (technologically) competent and having a style of use that distinguishes them" (2004, p. 83). This differentiation makes young people the contemporary protagonists of the new uses and functions of mobile phones. Our purpose is to analyze the images of young people in mobile ads in order to understand the uses, functions and technical characteristics of mobile telephones associated with their lifestyles in advertising discourse. To study the images of young people in the advertising context (in other words, to interpret the associated meanings in the different advertising texts analyzed in this paper), we use the basic theory of traditional semiotics. The contributions of Saussure (2000, signifier-signified), Peirce (icon, index and symbol), Umberto Eco (2000, image-coding information) and Roland Barthes (1994, denotation, connotation, level of ideology, the role of the image's anchor and relief) have contributed to the interpretation of each sign (images and text) in the different advertising texts.

To structure our analysis file with these authors, we have also followed the theoretical contributions of González Martín (1996), who, from the classification of the functions of advertising language established by George Péninou (Distinctive function, predicate function and implicated function), divides the semiotic study of advertising into three areas: syntax, semantics and pragmatics. This work is a semantic study of the advertising message. Semantics is the study of the relationship between the sign, the meaning and the reference. It is the study of the meanings present in the text, analyzing the meanings associated with the signs derived from productive resources (rhetorical operations), creativity and the operational rules of the signs within the text.

Finally, our work identifies the signs that act as symbols in the texts. Semiotics asserts that symbols are signs that contain a higher semantic category, and therefore, the correct interpretation will enable us to determine how the image of young people in the mobile telephone sector is constructed.

In this framework, the questions that guide our objectives and results are as follows:

• What meanings (connotations and denotations) are associated with the representation of the product (mobile phone) in the analyzed texts?
• What meanings (connotations and denotations) are associated with the brands present in the analyzed texts?
• What meanings (connotations and denotations) are most commonly associated with images of young people in the analyzed advertisements?
• Can we identify youth lifestyles in the analyzed texts?
• Can a youth lifestyle be detected in the analyzed texts that we can attribute to the new young consumer defined in the introduction to this work?

3   Methodology, Sample and Variables in the Studio

Our analysis offers results divided into two blocks. The first is to identify the most common signs, formal relations, and productive and creative resources, and the second is to study the semantics of the youth images in the ads.

To achieve our goal, we have developed a structured analysis file for twenty variables and recorded and analyzed it with SPSS software using descriptive analysis and frequency and contingency tables of two variables.

Our sample consists of 99 advertisements extracted from the two mainstream newspapers with the largest audiences in Spain: *El País* and *El Mundo*. To validate the audience data, we used the *EGM*[1] (*Estudio General de Medios*) data source prepared by the *AIMC* (*Asociación para la Invetigación de Medios en España*), which is the main source for measuring the print media audience in Spain. According to these data, the four most widely read newspapers in Spain are as follows:

---

[1]   The source of these data is the latest study published by EGM previous to submitting this work, which corresponds with data collected from February to November 2009.

1. *Marca*: 2.800.000 readers per day
2. *El País*: 2.081.000 readers per day
3. *El Mundo*: 1.309.000 readers per day
4. *As*: 1.306.000 readers per day

Although *Marca* is the newspaper with the largest readership in Spain, we decided not to analyze it. Both *As* and *Marca* are regular sports information papers, and we decided that this could invalidate the sample. Thus, our study centers on the analysis of images of young people appearing in mobile ads placed only in mainstream newspapers, and for this purpose, we used two national newspapers with wide audiences.

To select the advertisements and validate the representativeness of the sample, we established the following criteria:

1. Our investigation period includes the period of time from January 2, 2009, to December 25, 2009.
2. All copies of both newspapers in that period of time were viewed (698 newspapers in total).
3. We selected ads that featured mobile phones, mobile phone services, mobile phone companies, handset manufacturers or new uses for mobile telephones.
4. The frequency of the occurrence of an ad was not considered.
5. The ads were photographed and recorded, registering their dates and newspapers in the copies.
6. The final sample contained 99 advertisements.

The analysis file consisted of the following variables (Table 1):

Table 1: Variables

Registration	Each ad has been appointed to a four-digit code (0001)
Date	Date of paper in which advert is found
Support	1. *El País* 2. *El Mundo*
Product	Description for advertised product and its characteristics
Brand	Advertiser Trademark
Slogan	Register slogan
Size	Ad Size Format chart
Promise or benefit	Promise or psychological approach to consumer product offered. To identify the product promise, we asked the following questions: What tangible or intangible benefit does the consumer obtain if he buys the product or service? E.g. Discount. Examples: Low Price, Stay connected with friends all the time at low cost, others.
Idea	Creative Idea which describes promise/appeal of product
Manifesto	Narrative or artistic expression of ad
Literal Narrative	Story the ad tells
Business Strategy	Description of target advertising and target group it is addressing.
Age of the characters	1. Young People 2. Other. 3. Undetermined.
Social Status	High social class, media social class, low social class.
Sex	Male / Female
Description of the ethnic characteristics	Are ethnic features an important element in the construction of youth representation?
Description of work / professions	Is work / profession a feature to represent the image of young people?
Role of character in the ad	1. Is the target. 2. Is the spokesperson of the message. 3. Performing an action. 4. Receiving an action. 5. Is a mere spectator of the action. 6. It is the target group. 7. Other. When a person carries out more than one role, we register the main once.

Meanings to the image of young people	Associated values that describe characters' identity and image: We understand each image of youth as a sign that contributes to the creation of collective meaning of the message. We detected the denotative and connotative meanings of each image. We have interpreted the relationship between the meanings of each character's youthful image and the written text and the other signs present in the ad (brand and their meanings. With this relationship is an exchange of meanings is produced. Thus, we can detect the presence of elements of youth culture (music, fashion, activities, language, values) the image of the young that are associated with mobile phone. Similarly, we can identify the meanings (denotative and connotative) present in the mobile phone image and can determine when these are associated with the image of young people (technology, wireless, entertainment, communication). In this exchange, we choose the most outstanding significance. If this meaning is conditional on other meanings, then it is also recorded and analyzed to measure the frequency of individual significance and frequency of connection with other meanings.

4   Analysis: Results

Now that we have explained the methodology, the following section shows the quantitative results that structure our analysis based on the relationships among the study variables: advertiser, brand, product/service, promise and the meanings associated with the images of the youthful characters in the ads.

In section 4.1, we discuss the advertisers (network operator, manufacturer and sales point) that occurred most frequently in the analysis. We also provide the relationship among them, and finally, interpret the most frequent plots or persuasive strategies.

In section 4.2, we show the analysis of the images of the young people in the sample and show the relationship of the meanings most often associated with these images.

4.1   Mobile Telephony: Advertisers, Products and Applications.

According to our descriptive analysis (Figure 1), the advertiser that occurs the most often in the sample is Movistar, followed by Vodafone, Nokia, Yoigo, and

El Corte Inglés. However, this result should be reinterpreted. To ensure a clear interpretation of our results, we must recall the criteria for selecting our sample. We only post (consider) each ad once, and therefore, we do not measure the frequency of the occurrence of each brand or product. Instead, we consider different ads where each brand or product is present. That is, each ad will only be considered once. Another feature to take into account is that different versions of a single ad can belong to the same advertising campaign. We observe how Yoigo´s company has a frequency of 11 ads; however, they all belong to the same campaign, entirely in the month of August. As we shall see, they advertise the same product under a single business strategy and message but with different slogans. We must remember that the purpose of this study is to analyze youth identities proposed by the discourse of advertising in the mobile telephone sector and the proposed uses for mobile phones in advertising as a selling point. To measure the pressure or frequency of a brand, we preferred to observe its presence according its periods of emergence and the number of distinct ads that do not belong to the same campaign. According to the above criteria, we present the relationship of the most frequent advertisers in figure 1.

Except for Movistar, the frequency analysis shows that most ads promote or feature at least two brands: mobile manufacturers (Nokia, Motorola, Samsung, SonyEricsson), mobile telecommunication network companies (Movistar, Vodafone, Orange), and department stores or retail outlets (El Corte Inglés, Carrefour, Hipercor).

The most frequent advertisers are the telecommunications companies. The most frequent brand is Movistar, which is present in 15 exclusive poster ads, and in 19, it is present with other brands. Vodafone has only a frequency of 7 appearances as an exclusive advertiser but is present in 23 ads with other brands. Movistar and Vodafone compose more than 60% of the sample. The most frequent mobile manufacturers in advertisements are Nokia, Blackberry, Sony Ericsson, Motorola, LG and Samsung. El Corte Inglés is the department store with most outlets and ads.

The results of figure 1 show us the variety of products found in the analysis. We can observe how mobile advertising is not limited to promoting phones. On the contrary, mobile phone brands prefer to associate with telecommunications service companies and distributors or outlets to sell their products through various schemes aimed at different consumers.

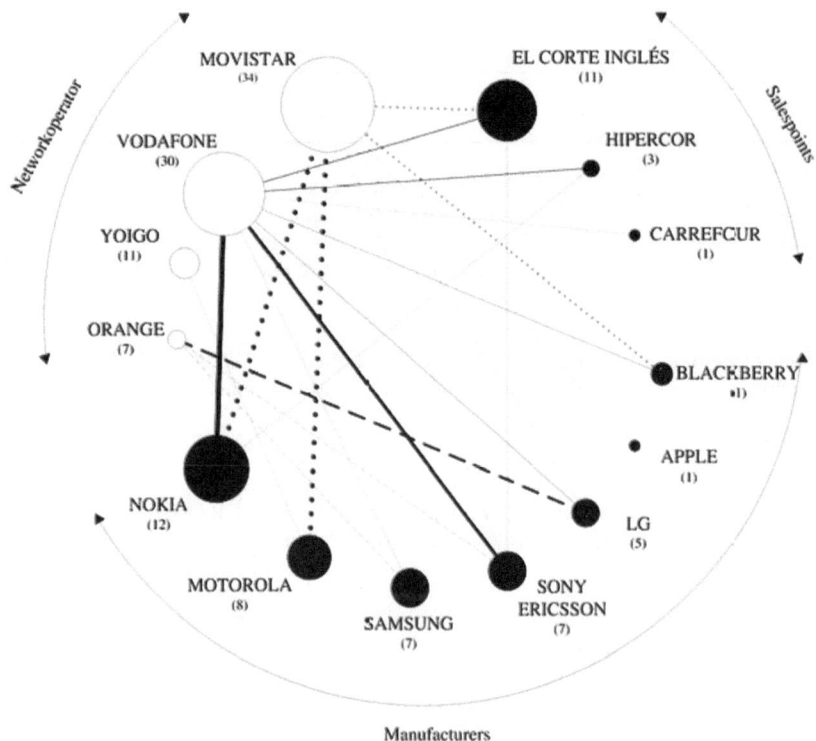

Figure 1: Frequency of advertisers and relationships among advertisers.

Table 2 illustrates the quantitative relationship among the following varia-bles: "product", "brand" (manufacturers, sales points, and network operators), "promise or benefit" and "targets." This relationship enables us to group promi-ses, types of arguments and persuasive strategies. This grouping helps us to more easily understand the "added values/ meanings" associated with mobile phone use as "advantages" and observe how the images show the presence of juveniles according to the values and meanings associated with the product. Therefore, we reduced the 99 advertising promises to 6 types of semantic arguments, which allows us to more easily analyze the meanings associated with images of the young people in the type of advertising that we study in section 4.2 of this work.

Table 2: Types of arguments or persuasive strategies.

Name	Product / Persuasive Strategy	Frequency
Type 1: Flat rate by service	Advertising that offers: Fixed price (FLAT RATE) or Mobile contract: To young people / By time / Call from abroad (roaming) / Browse Internet from mobile phone / To send SMS / Card Pack	21
Type 2: Mobile offer in "Department store"	Advertising that offers: Fixed price (FLAT RATE) or Mobile contract with mobile telephone in a department store.	13
Type 3 More flat rate mobile phone	Advertising that offers: Fixed price (FLAT RATE) or Mobile contract with mobile telephone.	36
Type 4 Manufacturer advertising	Advertising only of the manufacturers of mobile telephones.	8
Type 5 Sponsorship	Advertising or Corporate image and musical sponsorship	4
Type 6 New support for media consumption	The mobile telephone is used to watch television or to read newspapers. Many media outlets offer their content via mobile	4

We perceived six ways (Table 2) to offer products or trading strategies in our analysis. However, there are two main forms of mobile advertising. Thus, some advertising (*Type 1. Flat rate by service*) offers a specific contract with a tele-communications network company under a pricing strategy ("flat rate", "Pack card", "Roaming"). In other cases (*Type 2. Mobile offer in "Department store" and Type 3. More flat rate mobile phone*), the pricing strategy is accompanied by the promotion of a mobile phone. Type 2 determines the recruitment of a service to a particular retail outlet (Figure 2), while Type 3 is a service offer for a telecommunications network company and a mobile phone model (Figure 3).

Figure 2: Slogan "Move to Movistar at   *Figure 3:* Slogan "Talk with every-
El Corte Inglés"   body 24 hours a day per
(January 25, 2009, *El Mundo*)   1 € /day"
                                                                   (August 2, 2009, *El País*)

Services offered by mobile companies (Type 1) focus on the price of a call and the price of an SMS according to the type of consumer that the ad reaches. In one example, by advertising "Roaming", Vodafone offers a special price for young people who call from Europe to Spain. However, there are also offers for calls by professionals and the "self-employed", calls for a certain price at fixed hours in the day, prices per minute and fixed prices for browsing the Internet on mobile phones.

Type 3 (Figure 3) is the most common and uses price as a primary strategy in two ways: the service price and the price of the mobile phone. This marketing strategy is the most common and used by all of the mobile telecommunications companies (Movistar, Vodafone, Orange, Yoigo and Zeromóvil).

Type 4 (Figure 4) focuses exclusively on advertising by mobile phone manufacturers and shows how companies focus on providing the specific qualities of its products and not on their prices. Its frequency of appearance in this analysis is just five ads, because the purchase of a mobile phone in Spain is associated with the recruitment of phone service with a particular telecommunications network company. This type of advertising serves to generate demand among consumers for mobile phones but is determined by a call or SMS contract with a company.

Figure 4: Slogan "Life is Heavy Metal" (March 21, 2009, *El Mundo*)

A more detailed analysis of the graphic composition of the ads reveals the role of mobile phone images in the advertisements included in Types 2 and 3 when constructing the final meaning of the message. The construction of these types of ads is very simple yet sophisticated, according to the following structure: The text incorporates the essential information of the consumer promise offered by the telecommunications company, and the image of the mobile phone and its additions give a true value to the offer through the type of audience at which they are aimed and the consumer's previous knowledge of the product (telephone). In our view, the texts and the images of these examples pertain to the regime of denotative meaning. The text describes the characteristics of the advertising promise (flat rate, mobile gift), and the image complements the promise (the message). It conditions the meaning of the text. According to Péninou (1976, p. 184), one can speak of "display advertising" where the image emphasizes the importance of the text by repetition, which insists on showing the flat rate as a product detail.

If we observe the formats of the Vodafone advertising outlined above, then we notice that the mobile image is the centerpiece of all of the advertisements, and it is the promise of what the consumer will acquire if he accepts the terms of engagement with the telecommunications network company (first level of meaning or significance denoted).

Although we have classified the product in Type 3, the mobile phone is the true promise of the ad, and it is no longer only a product to communicate via voice at a fixed price.

If we examine the screens of all of the phones, then we can appreciate the different possibilities or specifications through the small icons that make up their menus. These become the first level of value denoted (to the price) and condition the consumers' perception, creating new signs that comply with the principle of the economy of language advertising. Therefore, to highlight its value as a camera, *mpx* is an abbreviation for "mega pixel camera", where in others, *MG* specifies the phone's ability to store data. Furthermore, some phones have a map showing the mobile phone's capability to function as *GPS* (Global Positioning System).

In the next examples, we show a series of advertisements that we think give new meaning to the mobile phone in the context of their primary or traditional functions (calling and sending short message service). These meanings are associated with the appeals of the new functions of cellular telephone services, such as watching television, playing music and messaging. When these meanings are associated with the mobile phone images in the ad text and are used to provide an arbitrary benefit to the consumer, we suggest that the sign serves as a symbol. "In semiotics, the term symbol is used in a special way to create any sign raising an arbitrary relationship between signifier and signified" (Hall 2007, p. 18). For example, if an ad promises that *the mobile phone brand "x" is connected to social networks*, this is a promise associated with the mobile phone's functions. However, when this same promise is arbitrarily associated with the appeals of "connectivity" or "friendship", we consider that the mobile phone ad text contains these meanings at the connoted level, and it serves as a symbol in this text. If the use of the mobile phone images as an arbitrary sign of "connectivity" or "friendship" is repeated with a high frequency, then we could state that mobile advertising works in certain contexts as a symbol of friendship or connectivity in discourse advertising in the mobile phone industry.

Figure 5: Slogan: "All you need it
is inside"
(January 9, *El Mundo*)

Figure 6: Slogan: "At this time we all
want the same" (December
13, 2009, *El País*)

Figure 7: Slogan: "Best TV with
Canal+ is in your phone"
(November 21, 2009, *El
Mundo*)

Figure 8: Slogan: New ways of
reading *El Mundo*
(January 7, 2009, *El
Mundo*)

In these ads, mobile phone photography is a symbol of communication and technology. Its meaning is broad and abstract, and it is no longer only a communication tool. The meanings of this symbol are, for example, entertainment, connectivity and mobility (Figure 5). It is the main protagonist of 50% of advertisements where no image reflects the potential consumer, but the features that make it the best tool are reflected. For example, it is depicted as the most desired Christmas gift (Figure 6), a radio, a portable computer, a sign of social distinction, an instrument of entertainment, a camera, a music player, a tool for browsing the Internet, a tool to chat, a messenger, a television (Figure 7), a new support for reading the news or the best tool to connect to social networks (Figure 8). These examples in figure 5 and figure 6 detail the need to offer the mobile phone as an object of desire that is necessarily associated with an offer initiated by telecommunications network company.

4.2 Youth Images and Mobile Phone Advertising.

Table 3 shows the frequency of juvenile characters in advertisements: almost 40% of ads show young people as characters in persuasive messages.

Table 3: Characters

Presence of young characters in the sample	Frequency
Ads with juvenile characters/adolescents	38
Ads with adult characters	7
Ads without characters/people	54
Total	99

After cross-checking these data with the "role"[2] variable, we note that, in virtually all mobile telephone ads (27 out of 35) with images of young people,

---

[2] We analyzed the relationship between the text and image of each ad to determine the "role" of the characters. We also consider the following items: 1. Character or characters name or describe the product in the first person, act as spokesperson for the product, and confirm or supplements the meaning of the text. Character or character name or describe the product in first person and identify themselves as consumers of the product. 3. Characters are described or named in the text as a target audience. 4. Characters act in the third person in the story and

these represent a *target group* at which the message is directed. In contrast to these results, the presence of adult characters is lower. Only seven ads in our sample have adult characters, all of which play the role of the consumer. Thus, based on our results, we can state that mobile phone advertising often uses images of young people and very few images of adults. We cannot empirically state the precise reasons for the use of images of young people in the advertising following the results of our analysis. However, if we rely on the contributions from José Luis León (2001), who, through a myth-hermeneutics analysis of advertising, reflected on the use of the mythological image of youth in advertising, we agree with his assertion that each representation reflects different intentions: e.g., to represent the target audience of the brand, to represent the product or service, to display the meanings and values of the brand through the image of the young, or to provide the meanings associated with the image of the young to the brand or the product.

In our study, we found that only one product, the Sony Ericsson Aino mobile phone, can be clearly identified as a product aimed at teenagers. The characters act as potential consumers in the advertisements, highlighting their preferences regarding mobile phones. Thus, music, video, sharing content with a PS3 and the camera are represented by youth jumping on large balls as a metaphor for the fun and excitement that teens prefer to enjoy with their friends. In our previous study, we agreed to use the term "youth advertising techno-culture" to signify a new era - a new temporary persuasive context - shaped by a set of messages aimed at the young, who act as the protagonists of technological change. One ad (November 2000) for the Sony Ericsson named these individuals *Generation WAP*, using term or extension used to denote old digital music files. The layout of this advertisement used the writing of short SMS messages as an expressive resource to define it. "You are gnration wap" thus identifies the generational culture in which they live. However, the mobile phone is a symbol that is desired for its technical characteristics and the identity and image personalization of the individual. It acts as an extension of individual identity, just like fashion and individual focal activities. Motorola ads, placed strategically on the pages of the EP3 weekly entertainment supplement of the newspaper *El País*, clearly reflect the intention of developing youth identities through the possession of a mobile phone. Other advertisements of this kind offer a mobile phone with camera to a vital personality, display a mobile phone for a conceited personality, or show a mobile phone with an image for a musical personality.

---

there is no sign identifying it as a target audience. 5. The role of character in the advertisement is ambiguous or just clarifies the meaning of the text. 6. The text appeals to the consumer through the "you". 7. The image uses signs of youth culture: music, aesthetics, recreation, education, standards, friendship, etc.).

Table 5 shows the results of the variables associated with the young characters' meaning, symbolism and age (to review the methodology, see table 1). These show us how the meaning of "fun" is used to configure the personalities and lifestyles prevailing in youth mobile advertising in 2009. We believe that there are nine cases in which fun is the primary meaning associated with youth. However, in all of these cases, "fun" is also associated with other qualifying meanings. In line with the theoretical contributions of Carles Feixa, the youth lifestyle can be understood as the symbolic manifestation of youth culture, "the set of tangible and intangible assets used by young people to express publicly their social identity, using bricolage techniques and homology, this is expressed through language, aesthetics, music, cultural creations and focus activities" (1998, p. 269). From this perspective, our analysis identifies juvenile images, and we will consider the cultural manifestations manipulated (by the advertiser) through the language of advertising. "In order to talk about youth we have to start asking ourselves: who applies it, to whom and for what?" (Criado, 1998, p. 36).

Thus, friendship, communication, love and music make up the framework of a fun and vital lifestyle. A *techno-cultural* lifestyle is centered around mobile communication and the social need of young individuals to relate to their own generation in their surrounding social and territorial context. These advertising characters do not faithfully recreate the ways that the young act. Instead, they use scenarios and preferences to generate a global advertising discourse in which the consumer does not literally see "youth" but creates a mental image associated with consumerism and the specific values of brand and/or product.

Table 4: Meanings associated with the young characters

Meanings associated with the characters	Frequency
Enjoyment associated with friendship, communication, love, culture or going out at weekends	9
Fashion	4
Music	3
Technology	3
Employment	3
Elite sports	2

Vitality	2
Pride	2
Consumption	1
Privacy	1
Solidarity	1
Variety	1
Travel	1

The newest and most prominent stage of this analysis involving young people has been created by ads that offer mobile social networks as a new value associated with the use of mobile phones. We say *"new"* because we clearly noticed in the second half of the year that social networks are featured as an added value in the sale of mobile phones. For example, one ad offers an urban and fugacious world entitled "your universe". This universe is cold and dehumanized, except for small images on the screen of a mobile phone and blurred human figures. The dehumanization reflects a lack of intent on the part of the advertiser to clearly reflect the consumer, expecting that the consumer will choose the product for its technological qualities.

Facebook advertising appears on the scene of Spanish mobile telephones through Nokia advertising by promoting a *hyper-individualized* identity. The identity proposed for the consumer is its *hyper-specialized* and *hyper-individualized* universe. The consumer then shapes the social meanings that give significance to the mobile phone identity, and, thus, mobile advertising uses *hyper-individualization* to promote a new style of young consumer.

Hyper-individualization requires that consumers feel that he or she is the one who creates the product, and not vice versa. Thus, they choose the context and their friends. Who to contact (and who not to contact) are finally described to us through the advertising concept used in advertisements, for example by Vodafone.

Under the headline "Your world and your friends always with you" of an advertisement by Sony Erisson, we can read the following copy: "*A friend wants to tell his latest adventure. Your old roommate has posted his photos from Australia and tonight there's a party at Andrew's house. All this is happening while you're on the street. Would you like to connect to Facebook on any website so as not be left out?*" This text can describe the consumer's lifestyle and the world (a collective universe of social networks) to which he belongs. A young and economically independent individual living alone can engage in leisure activities (such as travel) without any restrictive family ties. However, these activities no longer determine an identity because the "world" is no longer out there but, rather, is within a virtual world created through Facebook. However, paradoxically, the new "universe" present on a social networking site and offered from the point of view of a single concept (your world) actually feeds on the experiences from the outside world to which each individual necessarily belongs. In the new "virtual world" of advertising, there is a new backdrop in which young people have a *hyper-self-expression* with a pressing need to be publicly exposed and be connected.

5 Conclusions

Mobile advertising in Spain is determined by the investment of two major telecommunications network companies, Movistar and Vodafone. Therefore, most mobile advertising campaigns in this country involve using the services offered by these two companies, usually implying discounts for calls on phone contracts and discounts on the purchase of mobile phones. We found that mobile phone manufacturers advertise their new phone models and features (e.g., touch screen) in this type of advertising. In this context, young people are featured prominently, and more than 30% of the sample includes images of young people who act within ads as representatives of the target group to which the message is geared. Youth identity, styles and proposals are almost always related to concepts of fun and friendship, but the novelty of digital social networks that is proposed as a novel scenario for fun and friendship. We believe that the context of mobile advertising has changed. In our analysis, mobile phones and young people are projected in a semantic context constructed through meanings such as enjoyment, fashion, communication, friendship, love, music and technology. This relationship contributes to the idea that the image of the mobile phone in the ad is no longer only an iconic sign of its function. In the new context of advertising, the image of the mobile phone acts as a symbol with meanings arbitrarily associated with the image of young people through the signs of the youth culture that are projected in advertising. However, this change is not the main one detected in our analysis. In our opinion, the main change is the projected image of young people with the emergence of a new added value (functions) for mobile phones: social networking (e.g., Facebook and Tuenti) via mobile phone. Until then, the advertising images representing these meanings were signs of a youth culture located in a "real space". The advertising images in 2009 represent a "virtual space" of youth culture inside the mobile phone. Mobile phone technology has more attributes than ever before. The smart-phone advertising aimed at young people uses a mixed appeal; it promises new technological features and new uses attached to the values of youth culture. Now, mobile phone advertising targeted at young people has changed the message from "your world in your hand" to "your world within the mobile phone". The advertising reviewed in 2009 does not use images of young people who use their mobile phones in the traditional settings of youth culture. The mobile phone is no longer found against the backdrop of youth culture. Now, the stages of youth culture are "within the mobile phone".

References

Barthes, R. (1994). *The Semiotics Challenge*. Los Angeles: University of California Press.
Eco, U. (2000). *Semiotics and the philosophy of language*. Bloomington: Indiana University Press.
Feixa, C. (1998). *De jóvenes, bandas y tribus*. Barcelona: Ariel.
Gil, V. & Romero, F (2008). *Crossumer*. Barcelona: Gestión 2000.
González Martín, J. A. (1996*). Teoría general de la publicidad*. Madrid: Fondo de cultura económica.
Hall, S. (2007). *This Means This, This Means That: A User's Guide to Semiotics*. London: Laurence Kings Publishing.
León, J. L. (2001). *Mitoánalisis de la publicidad*. Barcelona: Ariel
Ling, R. (2004). *The Mobile Connection: The Cell Phone's Impact on Society*. San Francisco: Morgan Kaufman.
Lipovetsky, G. (2007). *La felicidad paradógica*. Barelona: Anagrama. (Original work published 2006)
Péninou, G. (1976). *Langage et image en publicité*. Barcelona: Gustavo Gili.
Saussure, F (2000). *Course in General Linguistics*. Illinois: Open Course Classic.
Thulin, E. & Vilhelmonson, B. (2008). Mobile phones: Transforming the everyday social communication practice of urban young. In S. Campbell and R. Ling (Eds.), *The Reconstruction of Space and Time: Mobile Communication Practices* (pp. 137-158). New Brunswick, London: Transaction Publishers.

Contributors

JULIEN FIGEAC, Ph.D., is Research Fellow at the Information Processing and Communication Laboratory, *Telecom ParisTech*, France. His research and publications focus on the uses of mobile TV and the forms of sociability generated by locative social media.

RALF HOHLFELD, Ph.D., is Professor for Communication Science at the University of Passau, Germany. He is Chair of the Department of Media Studies. His research and publications focus on the field of journalism research, communication theory, media performance, mobile communication, and media convergence.

ANNE JARRIGEON, Ph.D., is an anthropologist, doctor of communication sciences, and photographer. She is a Lecturer in urbanism at the University of Paris-Est and a member of the City – Mobility – Transport Laboratory, a shared research unit of *École des Ponts ParisTech*, the University of Marne-la-Vallée, and IFSTTAR, France. She conducts research combining approaches from ethnology, visual anthropology, and semiotics, and is particularly interested in the place of the body and images in urban practices.

VERONIKA KARNOWSKI, Ph.D., is Research Associate at the Institute for Communication Studies and Media Research in Ludwig Maximilians University of Munich, Germany. Her research and publications focus on diffusion processes, appropriation, and usage of new communication technologies, mobile communication, and web navigation and searching.

CORINNE MARTIN, Ph.D., is Assistant Professor at the Centre for Research on Mediations of the Paul Verlaine University of Metz, France. Her research and publications focus on the usage and appropriation of information and communication technologies (mobile phone, mobile internet) in their multiple dimensions: the economical, technical, social, gender, and cultural dimensions.

FRANK MÖLLER, Ph.D., is a Research Fellow at the Tampere Peace Research Institute, University of Tampere, Finland, a former editor of *Cooperation and Conflict*, and a member of the Finnish Center of Excellence in Political Thought and Conceptual Change, Research team Politics and the Arts. His recent work has been published in such journals as *Alternatives, Security Dialogue, Review of International Studies, Wissenschaft und Frieden,* and *Peace Review*.

MIGUEL ÁNGEL NICOLÁS OJEDA, Ph.D., is Assistant Professor at the San Antonio Catholic University of Murcia (UCAM), Spain. His research and publications focus on the study of advertising, young people, and their representation in the advertisement

THILO VON PAPE, Ph.D., is Research Associate at the Institute for Communication Studies at the Hohenheim University in Stuttgart, Germany. His main research interests lie in the fields of mobile communication and diffusion and appropriation of new information and communication technologies.

IREN SCHULZ, M.A., works as Research Assistant in the coordinating project of the DFG funded Priority Program "Mediatized Worlds" at the Centre for Media, Communication and Information Research (ZeMKI) at the University of Bremen, Germany. In 2011 she was awarded a doctorate (Dr. phil.) with a thesis about digital media and the change of socialization in adolescence.

CORNELIA WOLF, Dipl. Journ., is Research Assistant at the Chair for Computer Mediated Communication at the University of Passau, Germany. Her research and publications focus on the field of new forms of communication and technologies (namely mobile and crossmedia communication) as well as the research on media reception and effects.